Tectonics of Suspect Terranes

Mountain building and
continental growth

Topics in the Earth Sciences

SERIES EDITORS

T.H. van Andel
Stanford University

Peter J. Smith
The Open University

Titles available

1. **Radioactive Waste Disposal and Geology**
   Konrad K. Krauskopf

2. **Catastrophic Episodes in Earth History**
   Claude C. Albritton, Jr

3. **Tectonics of Suspect Terranes**
   Mountain building and continental growth
   David G. Howell

4. **Geology of Construction Materials**
   John E. Prentice

Tectonics of Suspect Terranes

Mountain building and continental growth

David G. Howell
U.S. Geological Survey

London   New York
CHAPMAN AND HALL

First published in 1989 by Chapman and Hall Ltd
11 New Fetter Lane, London EC4P 4EE
Published in the USA by Chapman and Hall
29, West 35th Street, New York, NY 10001

© 1989 David G. Howell

Typeset in 11/12 pt Bembo by
Thomson Press (India) Ltd, New Delhi, India
Printed in Great Britain by
St Edmundsbury Press Ltd
Bury St Edmunds, Suffolk

ISBN 0 412 30360 (HB)
ISBN 0 412 30370 (PB)

British Library Cataloguing in Publication Data

Howell, David G.
    Tectonics of suspect terranes: mountain
    building and continental growth
    1. Plate tectonics
    I. Title II. Series
    551.1'36

    ISBN 0–412–30360–4
    ISBN 0–412–30370–1 Pbk

Library of Congress Cataloging in Publication Data

Howell, D. G.
    Tectonics of suspect terranes: mountain building and continental
    growth/David G. Howell.
    p.   cm.—(Topics in the earth sciences)
    Bibliography: p.
    Includes index.
    ISBN 0–412–30360–1.–ISBN 0–412–30370–4 (pbk.)
    1. Geodynamics. 2. Orogeny. 3. Plate tectonics. 4. Continents.
    I. Title. II. Series.
    QE501.H733   1989
    551.1'39—dc19

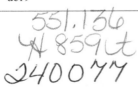

Contents

Series Foreword   vii
Preface   ix

1 Introduction   **1**

1.1 Continents   2
1.2 Oceans   6
1.3 Mountains   8
1.4 Thermally thickened crust   18
1.5 Continental growth   19
1.6 Conclusion   21

2 Plate tectonics: principal elements   **22**

2.1 Prologue   22
2.2 Plate tectonics   22
2.3 Conclusion   51

3 History of continental growth   **52**

3.1 Crustal recycling   52
3.2 A global budget   76
3.3 Conclusion   79

4 Suspect terranes   **80**

4.1 The rationale   80
4.2 Nomenclature   82
4.3 The making of terranes   89
4.4 Precambrian terranes   111
4.5 Conclusion   119

5 Kinematics measuring terrane displacements   **121**

5.1 Overview   121

5.2 Direct measurements   122
5.3 Magnetic lineations   124
5.4 Paleomagnetism   132
5.5 Paleontology   145
5.6 Fault-plane solutions   148
5.7 Conclusion   155

6   Mountain building and the shaping of continents   157

6.1 Overview   157
6.2 Taiwan to Timor   159
6.3 Himalaya and Tibet plateau   167
6.4 Africa–Europe collision   173
6.5 Cordillera of North America   180
6.6 The Andes   195
6.7 Conclusion   199

7   The strategy of a field geologist   200

7.1 The role of field geology   200
7.2 A field-mapping strategy   203
7.3 Conclusion   206

Glossary of terms   208
References   216
Index   225

Series foreword

Year by year the Earth sciences grow more diverse, with an inevitable increase in the degree to which rampant specialization isolates the practitioners of an ever larger number of subfields. An increasing emphasis on sophisticated mathematics, physics and chemistry as well as the use of advanced technology have set up barriers often impenetrable to the uninitiated. Ironically, the potential value of many specialities for other, often non-contiguous once has also increased. What is at the present time quiet, unseen work in a remote corner of our discipline, may tomorrow enhance, even revitalize some entirely different area.

The rising flood of research reports has drastically cut the time we have available for free reading. The enormous proliferation of journals expressly aimed at small, select audiences has raised the threshold of access to a large part of the literature so much that many of us are unable to cross it.

This, most would agree, is not only unfortunate but downright dangerous, limiting by sheer bulk of paper or difficulty of comprehension, the flow of information across the Earth sciences because, after all it is just one earth that we all study, and cross fertilization is the key to progress. If one knows where to obtain much needed data or inspiration, no effort is too great. It is when we remain unaware of its existence (perhaps even in the office next door) that stagnation soon sets in.

This series attempts to balance, at least to some degree, the growing deficit in the exchange of knowledge. The concise, modestly demanding books, thorough but easily read and referenced only to a level that permits more advanced pursuit will, we hope, introduce many of us to the varied interests and insights in the Earth of many others.

The series, of which the book forms a part, does not have a strict plan. The emergence and identification of timely subjects and the availability of thoughtful authors, guide more than design the list and order of topics. May they over the years break a path for us to new or little-known territories in the Earth sciences without doubting our intelligence, insulting our erudition or demanding excessive effort.

Tjeerd H. van Andel and Peter J. Smith
Series Editors

Preface

Prior to the theory of plate tectonics, disparate rock bodies of a continental margin were conceptually grouped into a profusion of geosynclinal models. The complexities of continental margins still remain, of course, but a sense of dynamic order flows from the new paradigm of plate tectonics. Modeling plate motions kinematically integrates the processes of seafloor spreading, subduction, and crustal rifting. This integration provides a rationale to link a host of rock bodies. Commonly, however, stratigraphic units that compose orogenic belts are fault-bounded and the geologic histories of neighboring units have contrasting elements; thus, their origin and close spatial linkage remain suspect. These fault-bounded units, called tectonostratigraphic terranes, may have formed elsewhere and subsequently moved into proximity to each other. The collage of terranes recognized in orogenic foldbelts may record a fundamental process affecting mountain building and continental growth. Understanding the stratigraphic, petrologic, and structural character of terranes, and how individual terranes relate to one another and to the neighboring continental regions, represents the basic precept of terrane analysis. There is nothing new or mysterious about this procedure. It differs, however, from geosynclinal modeling, and even from many of the classic plate-tectonic reconstructions, in the tacit understanding of the great mobility of the Earth's plates and the fragmentary nature of the crustal components. The various elements that now compose a foldbelt may have had diverse origins in widely spaced settings.

In most geological analyses one first maps geometric relationships, which provide a basis for tectonic reconstructions and a kinematic appraisal. From there, tectonicists can hypothesize about dynamic processes, the forces in the Earth that drive plate tectonics. Geologic processes operating today have probably been operating in the past. A facile application of this so-called uniformitarianistic principle could be misleading. In the time interval 3.8–0.6 Ga (giga-annum or 10^9 years ago), many lacunae are apparent in the rock record. Discrete events 200 to 300 million years (m.y.) in length may be over-represented in the rock record; for example, igneous rocks world-wide seem to cluster at 2.9–2.6, 1.9–1.7, and 1.1–0.9 Ga. Yet since 900 Ma (mega-annum or 10^6 years ago), the chronologic signature is more uniformly distributed.

Are there fundamental differences between old Precambrian and younger Phanerozoic igneous and tectonic processes? I suspect not. Viewing the Earth as a heat engine and plate tectonics as a cooling mechanism, there is no obvious reason why the effects of plate tectonics should have changed dramatically through time. Because of the nature of radioactive decay, the heat flux through the crust probably has diminished with time, but geophysical theoreticians are divided on what effects this may have had on rates of plate-tectonic processes.

Our problem in reconstructing the early history of continents may involve some kind of geologic dyslexia. The stories read from rocks seem to change with each decade as geologists develop new observational instruments and as they carry new hypotheses into the field. The objective of this book is to explain the rationale and concepts of terrane analysis, a new means by which to read rocks, especially those making up the complex array of strata that accumulate along active continental margins or that now crop out in orogenic foldbelts. Numerous examples will be drawn from around the Earth, particularly involving the past 200 m.y. of Earth history that followed the breakup of Pangea. With these examples providing a framework, I will venture into older terranes hoping to convince readers of the efficacy of plate tectonics and terrane analysis to understand mountain-building processes and the growth of continents since 3.8 Ga.

Some questions that are only now being answered include the following: Have continents grown since 2.5 Ga? How did (do) continents grow? What is continental freeboard and why is this concept important? What are the present rates of growth of oceanic crust, volcanic island arcs, and seamounts? How does the crust thicken beneath mountain systems? Does sediment and continental crust get subducted into the mantle? What is a tectonostratigraphic terrane and how does it differ from a terrain? What are the essential elements that distinguish geologic maps, tectonic maps, terrane maps, and paleogeographic maps, and how do they vary? What is the difference between the Wilson cycle (involving the opening and closing of ocean basins) and the terrane cycle (involving rifting, amalgamation, accretion, and dispersion)? How have ocean ridges changed configurations through time? Are terranes found only in California and Alaska? Is continental crust composed of flakes or is this a concept imposed by some geologists – in which case maybe the geologists are the flakes? Definitive or uncontroversial answers are not always provided, but I have striven to embrace the current thinking regarding these exciting topics. The approach is not always balanced; my personal persuasions are intended to stimulate the reader rather than to solicit converts.

I would like to thank the following who read either the book in its entirety or selected chapters, and whose many comments helped to improve both the clarity and the substance of the book: Tom Wiley, Roland von Huene, Jim Bischoff, Jack Vedder, Francois Roure, Claude Rangin, Holly Wagner, Susan Boundy-Sanders, Jerry van Andel, Jack Hillhouse, Rick Blakely, Joe Wooden, Duane Champion, A.M. Celāl Sengör, J. P. Swinchatt and especially George Moore who must have sharpened his pencil numerous times while correcting early versions of the text. Many of the ideas in this book were learned from friends and colleagues such as Davy Jones, Peter Coney, Paul Tapponier, Ken Hsu, Clark Blake, Jacques Bourgois, the reviewers mentioned above who did double duty, and my many colleagues of the Ocean Drilling Project's Tectonics Panel who shared with me their knowledge of the tectonics of the Earth. The book was written during a year of intellectual growth spent at the University of Pierre and Marie Curie in Paris, France; I will always cherish their warm hospitality and I will continue to be thankful to the US Geological Survey for providing this opportunity. And finally I must acknowledge my wife, Susan, who offered encouragements to keep writing, even with all the humming from the computer that resided in the bedroom of our small Parisian apartment.

1

Introduction

On what common ground to begin an analysis of the origin of mountains and how continents grow is difficult to know, because the character of rocks, and therefore the geology, varies so much from place to place. In any given region the geologic history commonly involves a complex sequence of interconnected events, often with gaps in the record. Our understanding of how the Earth works is enhanced by many recent advances in the subdisciplines of Earth science. Paleontology and isotopic geochemistry can now tell us more clearly and accurately about the ages of rocks and the conditions under which they formed. Seismic reflection and refraction images give us locally precise glimpses into the configuration of rock bodies within the crust and upper mantle. Petrologic studies provide additional exciting information about the lower crust and mantle. Paleomagnetic data can now record in greater detail the wanderings of crustal fragments. The integration of these and the many other interdisciplinary studies now available invigorates modern geology. Field geologists are no longer alone on a mountainous escarpment, as they are backed up by an entourage of specialists who can aid the field analysis with laboratory studies.

The intent of this book is to provide more insight into mountain-building phenomena and to relate this insight to the processes that have been responsible for forming the continents during the past 4 billion years (b.y.). The further back in time one goes, the less certain one is about many former conclusions. Nonetheless, the existing rock record and an understanding of relatively recent events and processes couched in terms of the plate-tectonic paradigm allow one a clear grasp of fundamentals involving the evolution of the Earth's crustal environment.

Since the nineteenth century, geologists have understood that basically two domains constitute the crust of the Earth, continental and oceanic crust – the so-called sialic and simatic crusts of many older textbooks. As recently as the late 1950s people thought that the more mafic and dense simatic crust of ocean basins represented the oldest crust. This notion contributed to the rationale for the ill-fated Moho drilling project because they believed that a single deep borehole through the strata above the oceanic crust would sample horizons going back to the very beginning of the rock record. Plate tectonics has drastically altered our thinking on this subject, for we now know that oceanic crust is not the oldest part of the Earth's crust, but rather it is the youngest. Even though $309 \times 10^6 \, km^2$ of the $510 \times 10^6 \, km^2$ of the Earth's surface is covered by oceanic crust, the average age of the world's oceanic crust is only 55 Ma, covering an age span from zero to 200 m.y. old.

The distribution of ages of oceanic crust has been mapped precisely from the record of magnetic lineations and confirmed by K/Ar dates of crustal samples collected during the past several decades of deep-ocean drilling (for example by the Deep Sea Drilling Program – DSDP-and the current Ocean Drilling Project–ODP). The consequence of this new information is startling, for if ocean floor is recycled on an average of every 110 m.y., in the relatively short time since the beginning of the Paleozoic Era the equivalent of five world oceans have been formed and four of them have been obliterated. As is discussed in greater detail below, some parts of continents are at least 3.8 Ga, and therefore, if oceanic crust has been recycling for this entire period of time, the Earth's oceanic crust has been formed and destroyed at least 34 times! That represents a recycling of approximately 7% of the Earth's mantle or an area of oceanic crust equal to $10.5 \times 10^9 \, km^2$ (nearly 21 times the surface area of the Earth). These are minimum figures. Continents have probably been gradually increasing in size, and thus ancient oceans would have been larger and the rate of recycling probably more rapid as a consequence of a hotter Earth. These topics are discussed in more detail in Chapters 2 and 3, but the general notion of a mobile crust is the underlying thesis of this book.

1.1 CONTINENTS

The current volume of continental material, estimated at $7.6 \times 10^9 \, km^3$, is only approximately 0.6% of the volume of the Earth; yet it has taken nearly 4 b.y. to form. In contrast, the oceanic crust has formed in only 110 m.y. on average, and the differentiation of the entire core–mantle subdivisions must have been completed within the first 100 m.y. of Earth history. Continental crust once formed, however, is quite resilient. It may be reconstituted or metamorphosed within the crustal domain, but because of its low density, large portions of continental crust are probably never recycled back into the mantle. This controversial topic is expanded in the third chapter.

The composition of the upper crust is roughly that of dacite, while the bulk composition of the entire crust is more andesitic (see Table 1.1). Although there has been an evolving enrichment of K_2O, U, and Th in the upper crust, the bulk crustal composition has remained remarkably

Table 1.1 Average composition of the upper crust and that inferred for the crust as a whole (bulk crust). Taken from Taylor and McLennan (1985)

	Upper crust (%)	*Bulk crust (%)*
SiO	66.4	58.0
TiO_2	0.06	0.8
Al_2O_3	15.5	18.0
FeO	4.6	7.5
MnO	0.1	–
MgO	2.0	3.5
CaO	3.8	7.5
Na_2O	3.5	3.5
K_2O	3.3	1.5
P_2O_5	0.2	–

uniform during its 4 b.y. of formation. The enrichment of K_2O, U, and Th probably represents progressive recycling and enhancement of the vertical compositional zonation that characterizes the crust – mafic components near the base and more siliceous components near the top.

The Earth is now considered to be approximately 4.6 b.y. old, even though no known rocks of that age lie at the surface. Meteorites have been dated at that age and gabbroic anorthosite from the lunar highlands of the Moon has been dated at about 4.5 Ga. The Canyon Diablo meteorite, which landed in the southwestern USA, contains no uranium; therefore, the ratios of the various lead isotopes in that meteorite provide a ratio of nonradiogenic lead that can be considered primordial and equivalent to the ratio that existed during the consolidation of the solar nebula. Deep-sea mud provides a globally homogenized sample of many of the crust-forming elements, including the U–Pb system. Thus, by knowing the decay constants of uranium isotopes to lead isotopes, and by knowing what the lead isotopic ratio was when consolidation in the solar system began, it has been possible to determine the approximate 4.6 b.y. duration of the Earth's evolution.

The search for the oldest rock has been under way ever since we had the technology to date rocks. Most, if not all, of the likely candidates have been dated. Not all of the surface area of the continents is exposed above sea level, but of the nearly 180×10^6 km^2 that is exposed, less than 3×10^6 km^2 is pristine, unaltered Archean rock. The remaining portions either have been created since the Archean or have been modified as a consequence of younger thermal and tectonic events. Rocks including both metasediments

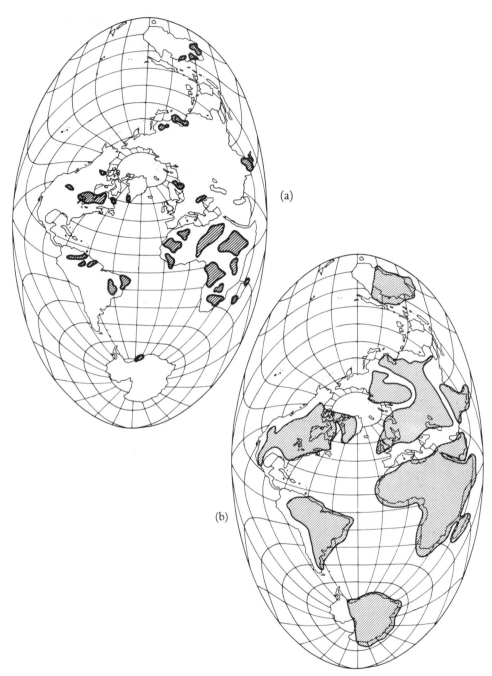

Figure 1.1 Mollweide transverse oblique map projections, scale averages approximately 1:250 M. (a) The general outcrop pattern of all known Archean strata, rocks older than 2.5 Ga. (b) The general outline of the Precambrian framework of continental blocks (older than about 570 Ma).

and deeper crustal igneous bodies with ages between 3.5 and 3.8 b.y. have been found in southwestern Australia, Labrador, Minnesota, Swaziland, and Greenland (see Figure 1.1). The oldest of these rocks is a sequence of metavolcanics, quartzites, and ironstones from Isukasia, Greenland: the Isua supercrustal sequence which Stephen Moorbath and his colleagues at Oxford University dated as 3.73 Ga. They also dated the Amitsoq Orthogneiss which intrudes the Isua supercrustals, and these various igneous bodies gave a range from 3.70 to 3.60 b.y. The basement on which the Isua strata were deposited has not been identified, and, of course, it must be older than 3.73 Ga.

In southwestern Australia, within the so-called Yilgarn craton, quartzites were deposited at approximately 3.1 Ga; but more importantly, within the detrital population are zircons that display a crystallographic habit indicating episodes of overgrowths. With the ion probe, one can zoom into the core region of each grain and date the oldest part. With this technique geologists at the Australian National University have determined zircon crystallization ages between 4.1 and 4.3 b.y. ago.

Both the Isua supercrustals and the Yilgarn detrital zircons indicate that continental growth was under way by at least 4 b.y. ago. But why have we not found any basement of these older ages? With a major effort devoted to searching for the oldest rock, something older than the Isua supercrustals and the Amitsoq Orthogneiss may be discovered; nonetheless, the chances are remote. During the early history of the Earth the incidence of meteoric impacts was far greater than today, probably by as much as 11 orders of magnitude. Many Earth scientists now believed that the Moon was created as the result of a collision between the Earth and a large asteroid-like body very early in the Earth's history (Figure 1.2). The lower density of the Moon, 0.6043 times that of the Earth, suggests that it is a portion of the Earth's mantle. Because samples from the Lunar Highlands are approximately 4.5 Ga, the Earth must have become stratified into the core–mantle partitions early in its history, and any crust that may have formed before this moon-forming collision would have been obliterated. The continued high incidence of impacts (Figure 1.3) would have pulverized any early bits of crust. Not until about 4 b.y. ago, after 600 million years of impacting bolides and the sweeping away of debris that cluttered the ecliptic, was the rate of crustal and lithospheric solidification greater than the destructive effects of meteoric bombardment. For these reasons, only a cryptic record remains of the early history of the continents.

The geologic record in the Proterozoic and certainly the Phanerozoic is much more complete, and therefore most of this book will concentrate on the past 2.5 b.y. of Earth history. The approach will be to analyze the ages and geologic configurations – the *geometric* analysis – and this permits an assessment of rates and movement histories – the *kinematic* analysis – which

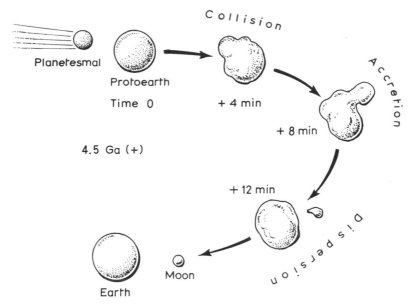

Figure 1.2 Schematic diagram portraying collision, accretion and dispersal phenomena in the 'collision–ejection' model to explain the Earth–Moon system. (Modified from Wood (1986) and Kipp and Melosh (1986).) The collision, accretion and dispersal tectonostratigraphic terranes associated with continental growth and mountain building are not nearly as dramatic, but as they form the topic of this book, it is hoped that they will prove to be equally exciting.

in turn provides a basis to understand possible forcing functions – the *dynamic* analysis. What is becoming increasingly accepted (even though as far back as the early 19th century insightful geologists were proposing similar models) is the extreme mobility of the Earth's crust. Fragments of this crust have been continuously dislodged from their points of origin and subsequently transported into exotic settings. These fragments are called tectonostratigraphic terranes, and with an understanding of the distribution and character of these terranes, of the rates of geologic phenomena, and of the kinds of forcing functions that accompany plate tectonics, we can comprehend the origin of mountain belts and the formation of continents.

1.2 OCEANS

Oceans play a key role in mountain- and continent-forming processes even though the rock record of the oceanic crust is rarely preserved for more

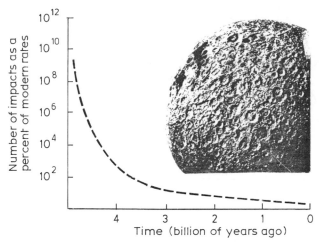

Figure 1.3 Graph depicting the estimated frequency of cratering on the Earth's surface due to collisions with planetesimals. These data are extrapolated from similar data determined for the Moon, where the record of impacts has been preserved. Note the rapid drop off in the incidence of collision by nearly ten orders of magnitude during the first 600 m.y. of Earth history. This was a period of sweeping up the clutter within the ecliptic of the newly formed solar system. Presumably one of these collisions was with a Mars-sized object, which resulted in the formation of the Moon, about 4.5 Ga.

than a few hundred m.y. Oceanic crust forms along the crest of the so-called 'midocean' ridges, although only in the Atlantic does the ridge system occupy a medial position. Today's oceans contain 56 000 km of globe-girdling ridges. Along the crest of this ridge system, molten rock erupts at about 1200° C and cools to form a 1–2 km thick cap of basalt. The magma below becomes chemically stratified, and forms an intermediate layer 3–4 km thick constituting a system of dikes and sills and solidified gabbroic magma chambers with minor amounts of more granitic material, and a lower layer commonly about 2 km thick of newly recrystallized ultramafic rock. These mafic (simatic) layers move away from the ridge crest at varying velocities, < 1 to 10 cm/yr in today's oceans, and ultimately descend back into the mantle along subduction zones. Thus, oceanic crust is viewed as the upper layer of a giant conveyer belt (Figure 1.4). The tectonic interactions along continental margins and the role that the conveying system of oceanic lithosphere plays in island-arc volcanism and tectonic accretionary processes are principal topics to be discussed later in this book.

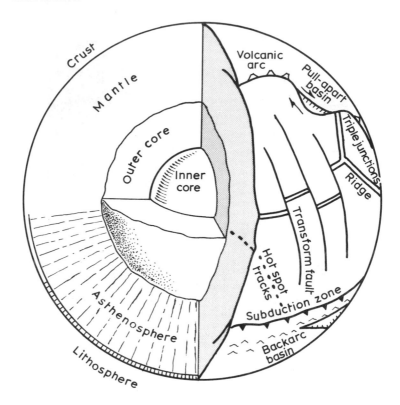

Figure 1.4 Schematic diagram displaying the principal feature of the lithosphere, the rigid 50 to 150 km thick plates that incessantly move about on the Earth's surface, born along the 56 000 km of spreading-ridge systems and consumed along the 36 000 km of subduction zones.

1.3 MOUNTAINS

In our effort to understand how mountains form, we must keep in check our instincts to exaggerate the dimensions of the seemingly rough outer skin of the Earth. Illustrations, maps, and models commonly exaggerate the vertical scale of landforms in order to highlight their occurrences or provide room to display internal features. Yet with an accurate rendering of the relief of the Earth's surface, on a standard desktop globe, one could not feel the topographic roughness of the highest mountains or the depressions of the world's deepest trenches. The elevation of a satellite revolving 200 km overhead would only be the thickness of a small coin above the surface of a desktop globe. Comprehending the vertical dimension of geologic units and the time of duration and rates of formation of these features is difficult. As you progress through this book try to visualize the actual dimensions of features despite the vertically exaggerated renderings of diagrams. Picture the surface of the Earth as you might see it out of the window during a transcontinental airflight, rather than how it seems to be from a vista point along a mountain trail.

The present average elevation of the Earth's surface above sea level is 780 m; this is higher than the Phanerozoic average (approximately 600 m) because of sea water currently housed in the polar ice caps. Locally, the rocks have been piled, buckled, or thermally driven up to 10 000 m above the surface of the sea. Mountain ranges can form quite rapidly. Maximum uplift rates over 1 cm/yr, that is 10 km/m.y., have been measured in places along the Southern Alps of New Zealand and in the Coast Ranges of California. Most of the high-range topography in the western USA is less than 15 Ma. The Alaska Range, one of the highest in the world, formed within the past 5 m.y. The area of South Island of New Zealand just 20 m.y. ago was covered by the sea, and since then has experienced as much as 30 km of uplift. The effects of erosion have almost kept space with uplift, as Mt Cook, the highest point in the Southern Alps of New Zealand, is only 3.3 km high. The opposing forces of constructional uplift and destructional erosion are constantly at work shaping our landscapes. For an area to remain at high relief requires that continual uplift occur, yet dynamic configurations of the Earth's crust are rarely stable for periods exceeding 25 m.y.

Mountain ranges reflect four basic configurations: tensional, compressional, transcurrent, and thermal. Figure 1.5 schematically portrays these.

Tensional mountain systems

In areas where the crust is under tensional stress, the crust thins by normal faulting, which results in an array of 'tipped blocks' and a basin-and-range physiography. The tensional forces are generally a consequence of gravitational pull along the flanks of an elevated region. High thermal anomalies may have uplifted an area to above-average elevation thereby providing a setting where tensional forces will prevail (Figure 1.6); alternatively, elevated areas that fail under tensional forces may be the consequence of crustal thickening, formed in an earlier regime of compression (Figure 1.7).

The early stages of rifting develop into a horst and graben structural setting (elevated ridges or mountains with intervening valleys). In instances where the initial crust was particularly thick, a 'ridge' may be a massive mountain range with an imposing range-front escarpment, for example,

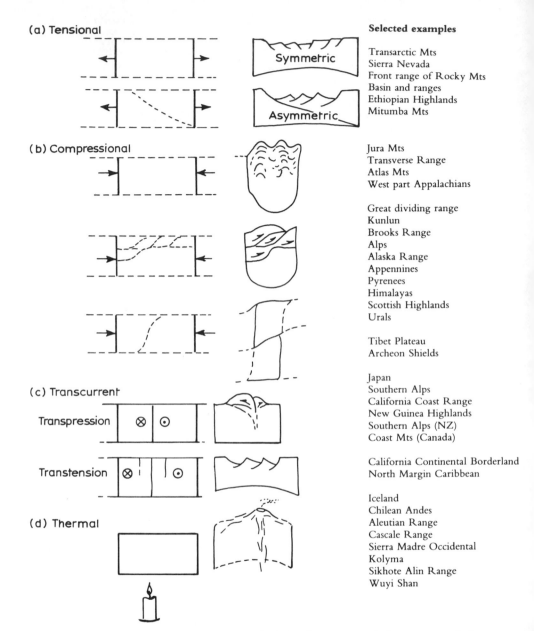

Selected examples

(a) Tensional

Symmetric

Asymmetric

Transarctic Mts
Sierra Nevada
Front range of Rocky Mts
Basin and ranges
Ethiopian Highlands
Mitumba Mts

(b) Compressional

Jura Mts
Transverse Range
Atlas Mts
West part Appalachians

Great dividing range
Kunlun
Brooks Range
Alps
Alaska Range
Appennines
Pyrenees
Himalayas
Scottish Highlands
Urals

Tibet Plateau
Archeon Shields

Japan
Southern Alps
California Coast Range
New Guinea Highlands
Southern Alps (NZ)
Coast Mts (Canada)

(c) Transcurrent

Transpression

Transtension

California Continental Borderland
North Margin Caribbean

Iceland
Chilean Andes
Aleutian Range
Cascale Range
Sierra Madre Occidental
Kolyma
Sikhote Alin Range
Wuyi Shan

(d) Thermal

Figure 1.5 (a) General classification for the disruption of the crust and the formation of mountains. As most mountain systems evince compound relations of more than one of these crustal thickening or thinning processes, the examples are therefore somewhat arbitrary. Either the principal or most recent deformational style has been chosen. Commonly, the present topographic expression of a mountain system may reflect one style of deformation, whereas the crustal thickening process(es) relates to a separate set of circumstances. Chapter 6 provides expanded discussion on this topic.

Figure 1.5 (b) Index map of mountain systems.

the east escarpments of the Transantarctic Mountains, Antarctica, the Sierra Nevada of California, or the Rocky Mountains of Colorado. When crustal stretching continues, an ocean basin may form, for like chilled taffy, continental crust can stretch only so far before it breaks. Approximately

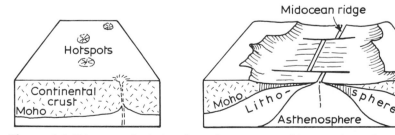

Figure 1.6 Diagram depicting how hotspots weaken continental crust. The elevation of the asthenosphere results in continental crustal attenuation, seafloor spreading, and the formation of an ocean basin.

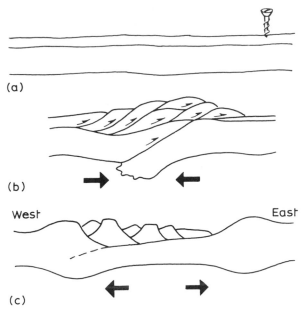

(a)

(b)

West   East

(c)

Figure 1.7 Compressional tectonics and associated plutonism (so-called classic Laramide orogeny) thickened the crust of western USA during the Cretaceous and early Tertiary. With the cessation of compression, the thickened crust collapsed under gravitational body forces. The configuration of attenuation shown in (c) is an alternative to the mode shown in Figure 1.8(c).

200 m.y. ago, the regions that now border the Atlantic Ocean were characterized by such a physiography; the Connecticut River valley and submarine depressions in the North Sea are remnants of the down-dropped grabens; many other grabens lie buried beneath sediment of the continental shelves along the two sides of the Atlantic. The regions of high relief have been worn down by erosional effects during the past 200 m.y. Under a tensional regime, where an ocean basin is formed some crustal blocks may become detached from the continent and be transported on the conveyer belt of the ocean crust. Examples include some of the submerged, but high-standing, plateaus of the ocean, particularly around the ancient margins of Gondwana: Exmouth, Agulhas, Kerguelen, and Lord Howe Rise. More discussion about these features follows in subsequent chapters because of their important role in accretion tectonics.

   The Basin and Range province of western North America is a modern example of a region where tensional forces are still active, and basaltic eruptions in the Rio Grande rift carry an oceanic geochemical signature possibly foretelling the future location of a new ocean basin. Tensional stresses in the Basin and Range may be due to a thermal response (Figure 1.8). Some geologists speculate that this could be a result of western

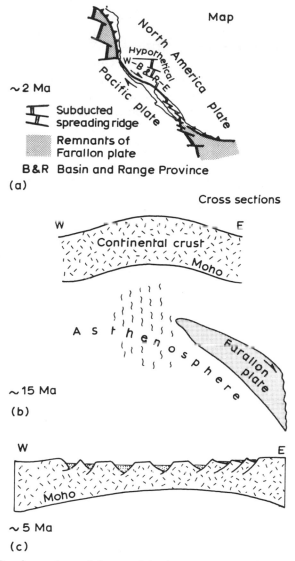

Figure 1.8 Geodynamic model to explain the attenuation of continental crust in the Basin and Range province of western USA. The northward projection of the East Pacific Rise below the crust (a) is envisaged as providing a thermal anomaly that elevates the crust resulting in gravitation collapse. (b) and (c) provide an alternative model for the cause of extension to the one shown in Figure 1.7.

North America overriding a segment of the East Pacific Rise spreading system. Alternatively, however, this crustal stretching may have arisen as a consequence of the relaxation of the forces that originally caused crustal thickening during the episodes of thrust faulting (Figure 1.7). Support for the latter supposition is found in some old fold-and-thrust mountain systems where gravity data indicate the absence of crustal roots – seismic-reflection images from these lower crustal regions are opaque, showing none of the thrust faults that presumably accompanied the convergent tectonics. These observations are consistent with the idea of lateral movement of crustal material away from the thickened region. In the upper crust, this process is expressed by brittle extensional faulting, while in the lower crust ductile flow predominates.

Two types of crustal fracturing can occur in a region of tension. Several ODP legs, scheduled for the 1990s, hope to identify specific areas typifying these two modes of crustal spreading. In one instance the lower part of the crust thins by horizontal stretching in what is called 'pure shear'. The upper 10–15 km of the crust is more brittle and therefore when stretched becomes thin by breaking into a series of blocks that rotate and collapse on one another. This is one form of what is called 'simple shear'. In the second mode of simple shear, crustal thinning occurs by breaking the crust along a low-angle surface that propagates through most, if not all, of the crust. As spreading continues, blocks above the *decollement* surface calve into a thinned zone. When this occurs during the breakup of a continent, the conjugate margins of the resultant ocean are nonsymmetric about the region of rifting, whereas in the first instance conjugate ocean margins have a nearly symmetrical structural fabric (see Figure 1.5(a)). The importance of this in oil exploration strategies is now beginning to be recognized.

Compressional mountain systems

In areas of compressional stress the crust shortens and thickens into a mountainous physiography. The thickening can occur in three styles (folding, thrust nappes, and doubling of crust), although all three normally play a role in any given compressional regime. In instances of *pure shear* the thickening is perpendicular to the principal shortening direction, and the thickening results from folding, with only minor faulting in the upper crust. For the upper crust to fold and shorten, it must be decoupled from the crust below. Crumpled crust such as the Jura Mountains is detached from the crust below, lying above a major decollement, in this instance a *blind thrust* in structural parlance – as in buried, out of sight. Approximately the upper 10 km of crust folds like a rug being pushed against a wall; the buried decollement is equivalent to the rug–floor interface. The primary structures accompanying this tectonic style are synclines and anticlines, and the amount of shortening determines the tightness of folding within these structures (Figure 1.5).

Numerous examples exist where thrusting is the principal mode of crustal thickening (Figure 1.5). The ratio of the amount of horizontal shortening to the vertical thickening depends upon whether or not basement strata are involved in the thrusting. The so-called *thin-skinned* situations are instances where the covering strata are decoupled from the basement, and supracrustal strata are stacked into thrust sheets above the basal detachment or decollement. In the European Alps, layers of salt and gypsum or hydrostatically overpressured shale provide surfaces along which strata can move, almost unencumbered by friction. In the Brooks Range of northern Alaska as much as 500 km of shortening has been accommodated by the stacking of thrust sheets that are commonly only 0.5–1.0 km thick. A 5 km deep exploratory well, that spudded in strata near the crest of this range, penetrated the same stratigraphic interval seven times. But in both of these cases, in order to explain the amount of crustal thickening, the basement at deeper levels must have somehow become involved in the thrusting. If the Alps are taken as an example, two styles of basement-involved thrusting are noted. In some occurrences nappes of crystalline basement show little internal tectonic strain. These lumps of granite and gneiss (*massifs*) are commonly called rigid basement to contrast with mobile basement that has been highly strained during the episode of tectonism.

Elsewhere, piles of sediment offscraped from the downgoing slab in subduction zones thicken as a consequence of thrust faulting and form an accretionary wedge. The basal decollement is generally within the thin horizon of pelagic mud that lies above the basaltic cap of the ocean crust. This horizon of detachment has been penetrated by a borehole in the Lesser Antilles arc system during Leg 112 of the ODP. In this zone, the hydrostatic pressure was found to equal the lithostatic pressure; thus, the layers of strata above the decollement are virtually floating.

Where large regions of crust approach a thickness of 70 km – nearly twice the average continental crustal thickness – some geologists have suggested that this represents a doubling-up of crust. The Tibetan Plateau and the Altiplano of the Peruvian Andes are two cases illustrative of this model; and in all fold and thrust belts accompanied by extensive amounts (generally > 50 km) of crustal shortening, the lower crust is probably involved in thrusting. Many of the Precambrian shield regions of the world have rock exposed at the surface with mineral phases indicating crystallization depths of 25 to 30 km. Because they are now at the surface, and because the crust below is still 35 to 45 km thick, these regions too may represent areas where the crust was doubled in thickness. However, this model of one crustal layer overriding another crustal layer is possibly too simple. Where deep seismic data (to depths of 30–40 km) are present, such as in the Pyrenees, the thickened crust represents a complex pattern of thin-skinned and thick-skinned thrusting, as well as a local doubling of the lower crust.

Transcurrent mountain systems

The orientations of dynamic forces stemming from plate tectonic processes range from orthogonal to parallel to the boundary for any given piece of crust. In the examples discussed above, a major component of the forces has been oriented at a high angle to the deforming crust. But plates also can slide past one another in a purely translational mode. In this mode, slight changes in plate motions or areas of small bends along transcurrent faults can cause local regions of either tension or compression, called *transtensional* or *transpressional* regimes respectively.

In transtensional regimes, the crust thins and pull-apart basins will form. Earthquake focal mechanisms along the Anatolian fault system in Turkey and faults along the north margin of Venezuela and across the southern California borderland reveal that both normal faulting and pulling apart of the crust have occurred in what is otherwise a strike-slip domain. Large basins such as the Sea of Japan or the South China Sea may also have formed as a result of transtensional tectonics.

Transpression may also occur locally along transcurrent fault systems. In these instances the crust is wrinkled into *en echelon* ridges and valleys subparallel to the strike of the fault. Many of the mountain ranges in Japan, Turkey, east Asia, coastal California, and along the northern and southern margins of the Caribbean Sea owe their existence to this phenomena. Where compressional stress affects an entire transcurrent fault system, such as along the Alpine fault of New Zealand, earlier formed basins evert, and a broad region of uplift forms a major mountain system. *Dispersion tectonics* is the long-term effect of both transtension, the ripping away of crustal pieces, and transpression, the accretion of pieces that have slid along the transcurrent fault system.

Thermal welts

Regions of the crust that lie above mantle upwelling systems generally have a high thermal gradient; the crust will rise due to the lower density attendant on the heating and the crust will thicken due to magmatic instrusions. Above subduction zones, magma rises into the lower crust, probably because of the frictional heating and emplacement of water into the mantle. The rising magma stimulates further melting and fractionation of the more volatile melts that rise to the surface creating explosive volcanism and forming chains of volcanic edifices. The island arcs of the Pacific and the Andes of South America are examples of these processes, and these are the principal regions where new continental material is being generated, at an average rate, as measured over the past 100 m.y., of approximately 1 km^3/yr (or nearly 32 m^3/sec). This topic is expanded in Chapter 3.

Deep within the mantle, possibly arising from the core–mantle interface, are relatively small, extraordinarily hot regions. These 'hotspots' seem to create plumes that rise toward the crust. Melting of a basaltic fraction occurs at a depth of about 100 km and jets of magma rise to the surface creating enormous shield volcanoes. The Hawaiian, Canary, and Tahitian islands, Iceland, and Mount Kilimanjaro are examples of this phenomenon. Because these hotspots are spatially stable within the deep mantle for intervals of 50 to 150 m.y. in duration, and because the crustal plates are shifting, the hotspot-generated volcanoes commonly form long linear chains with active eruptions at the present position of the hotspot, and progressively older volcanoes backward along the chain. The long Hawaiian–Emperor submarine mountain chain is particularly illustrative. This locus of eruptions provides one of the techniques to reconstruct ancient plate motions. The prominent bend in the Hawaiian–Emperor chain marks a major shifting of direction of the Pacific Plate. The volcanoes in the vicinity of this bend have been dated at 43 Ma, which is thought to relate to the approximate time that the continental crust of India impacted with Asia. Could this collision have caused a reorientation of the major plates of the world?

Because Iceland is situated directly on a spreading ridge, the eruptions do not result in a trail of volcanic mountains; instead, the long-term concentration of magmatism has resulted in an anomalously thickened piece of oceanic crust. Perhaps some of the oceanic plateaus such as Minihiki, Hess, Shatsky, or Ontong Java owe their existence to this phenomenon. More will be said about this in later chapters.

A hotspot lying below continental crust may create a chain of volcanoes if the crust is moving relative to the thermal plume. The Valley of Ten Thousand Smokes in the Snake River Plain and the thermally active part of nearby Yellowstone National Park, USA, are examples of this. If the continent is relatively stable, propagating rifts may form that could progress into full-scale crustal breakup. The East African rift system illustrates this and portends the eventual breakup of eastern Africa into smaller continental fragments. With continued rifting, the hotspots become localized along the midocean ridges. As many as 21 major hotspots are currently active; 14 are on or next to a midocean-spreading ridge, and of these 7 lie along the Mid–Atlantic ridge system.

1.4 THERMALLY THICKENED CRUST

One final category of crustal thickening must be considered in any continental-growth scenario. During the Proterozoic, from 1 to 1.5 b.y. ago, owing to a thermal anomaly, an enormous volume of magma was added to the crust of North America in an episode that was not associated with tectonism (Figure 1.9). Another huge igneous event, this time in the oceanic domain, occurred during Early Cretaceous time, approximately 110 to 90 m.y. ago. Throughout much of the Pacific, and also on the Caribbean plate, a thick layer of basalt is interstratified between Lower and Upper Cretaceous deep-marine sedimentary layers. An area of the western Pacific that encompasses these basaltic episodes seems to radiate outward from the Ontong Java Plateau. The volume of each of these igneous outpourings is at least an order of magnitude larger than any known igneous event in Tertiary time. What caused these events – intrusions stemming from impacting bolides, eruptions from the Earth's core, or something unknown and entirely different? How often have such events occurred? Were they more frequent during initial episodes of crustal

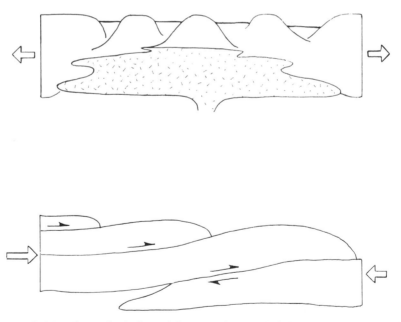

Figure 1.9 Two hypothetical models to explain crustal thickening processes. In one, thickening is a simple tectonic response to compression and shortening, whereas in the other, the crust first dilates in a tensional regime and then thickens as a result of intrusions, principally within the lower crust.

growth in the Early Archean? Answers to these questions await further research, but their significance is relevant to all discussions involving continental growth.

1.5 CONTINENTAL GROWTH

The growth history of continents remains a controversial topic. Figure 1.10 is a sampling of various growth-rate scenarios that are popular in today's literature. The subject is particularly confusing because continents are composed of recycled crustal rock in addition to first-generation mantle derivatives. The application of some isotopic dating systems, particularly Sm/Nd, is helping to solve this problem, as these data can provide times at which the host rock hatched from the mantle, regardless of all subsequent crustal processes (such as weathering, metamorphism, or melting and recrystallization). A collection of these kinds of data from a cratonal area can provide the percentage of rock that differentiated from the mantle at various times. As an example, the Churchill province of the Canadian shield underwent a tectono-thermal event at 1.9 – 1.7 b.y. ago, which reset many of the radiogenic clocks; yet Sm/Nd (which is unaffected by such events) shows that 95% of the rock was differentiated from the mantle over 2.7 m.y. ago (see Figure 1.11). This should be contrasted with the midcontinent region of North America where again the rocks indicate a

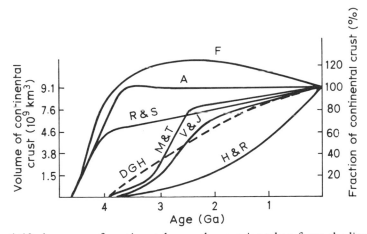

Figure 1.10 An array of continental growth scenarios taken from the literature. F = Fyfe (1978), A = Armstrong (1981), R & S = Reymer and Schubert (1984), M & T = McLennan and Taylor (1982), V & J = Veizer and Jansen (1979), H & R = Hurley and Rand (1969). DGH is the growth history model following the thesis of this book.

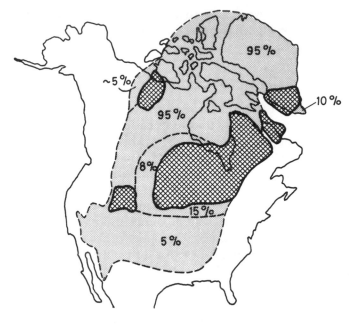

95% Percentage of Archaean rock admixed with crust
     affected by 2.0-1.7 orogenies

     100 % Archaean crust

Figure 1.11 Diagram of North America and Greenland showing the general outline of rocks older than 1.7 Ga specifying the percentage of Archean crust within specific domains. These data are based on Sm/Nd analyses. More data of this type are required before better agreement on the growth history of continents can be achieved. (Diagram modified from Pachett and Arndt (1986))

thermal event between 1.9 and 1.7 b.y. ago, but in this instance the crystallization represents a first cycle out of the mantle.

More data such as those just described are needed before agreement can be reached on the history and rates of continental growth. Chapter 3 reviews this general topic and provides a model of continental growth that follows line DGH of Figure 1.10. This model represents a blending of existing geochemical data with concepts of tectonic processes in order to reconstruct the evolution of plate-tectonic settings. It describes a growth rate that has been gradually decelerating as a consequence of the cooling of the Earth.

1.6 CONCLUSION

The fundamental thesis of this book is simple. New continental crust is formed in regions of thermal welts, principally island arcs above subduction zones and oceanic islands above hotspots. Continental growth occurs when these bodies are tectonically attached to the margins of an already existing continent. Compressional forces, associated with accretion tectonics, thicken the crust and are responsible for forming most mountain systems. Tensional forces thin the crust and reshape the continents; the rifted pieces of continent inevitably become accreted elsewhere. Thus, the continents of today represent a collage of disparate crustal fragments (tectonostratigraphic terranes) that are in a continual state of flux.

2

Plate tectonics: principal elements

2.1 PROLOGUE

The modern concepts of plate tectonics were formulated rapidly into a working hypothesis in the early to middle 1960s. Elements of the modern paradigm had been proposed since the early part of this century, but the ideas were too revolutionary at that time and did not take hold for want of a unifying principle. Despite the supporting evidence for continental drift, it was not widely accepted as a viable tectonic model because it seemed mechanically impossible. Isostasy required a sialic continent to have a root, in contrast to the simatic ocean crust. Seismic-refraction observations confirmed the evidence from gravity that continental crust is commonly 35 to 50 km thick, whereas oceanic crust is generally only about 6 to 8 km thick. How could continents with such a deep keel plow through the rigid oceanic crust especially because granite is fundamentally weaker than basalt. Many Earth scientists were left in a quandary of apparently conflicting evidence. As recently as the middle 1960s, most universities dealt pedagogically with continental drift in the way that universities today deal with the expanding-Earth model. A tectonic mechanism, a dynamic model, was needed before a significant part of the geologic community would accept the geometric reconstructions and the various kinematic scenarios promoted by the advocates of continental drift.

2.2 PLATE TECTONICS

Seafloor spreading

Harry Hess is universally accepted as the originator of the idea of seafloor spreading although it was his colleague, Robert Dietz, who actually coined the term in 1961, the year before the publication of Hess's seminal paper that set the stage for the birth of plate tectonics. What may have allowed Hess to make his great inductive leap was new information regarding the topography of the seafloor. Echo-sounding systems, developed during

World War II to map bathymetry in war zones, were used subsequently during the 1950s to chart the world's oceans. A startling revelation was the discovery of midocean ridges. Hess reasoned that these ridge systems define the locus of points where convection cells rise upward and roll over. The vertical swelling of the mantle produces the positive relief evinced by the ridges, and the lateral flow of the mantle convection propels the crust outward. New oceanic crust is formed in the plume of the upward-moving crust, and in order to maintain a uniform surface area for the Earth, crust must also be consumed, presumably evinced by the Wadati–Benioff zones★ of deep earthquakes. The geometry of seafloor spreading proposed by Hess is essentially correct, but, as will be discussed, his mechanism for the driving force of plate tectonics, involving the coupling of convection cells in the mantle to the base of the overriding plates, has been modified.

To validate the seafloor-spreading hypothesis, an independent test was required to confirm crustal spreading along the midocean ridges. Coincidental discoveries converged on the problem. Interest in paleomagnetism provided evidence that the Earth's magnetic field flips back and forth in its north and south orientations (normal and reversed polarities). During the same time, off the western coast of North America, a systematic magnetometer survey was made in concert with a bathymetric survey by researchers at the Scripps Institution of Oceanography. They discovered positive and negative magnetic anomalies that did not correspond to the bathymetric relief of the oceanic basement. They recognized that some unknown mechanism, associated with basaltic intrusion, had produced this linear magnetic-anomaly pattern. The magnetic lineations were also observed to be offset along fracture zones that trended across the magnetic stripes. From these data other geologists reasoned that the pattern of magnetic anomalies represented discrete intervals of either reverse or normal magnetic polarity, and these field reversals must be recording discrete episodes of basaltic intrusion. Furthermore, as it became known that these reversals are evident worldwide in ocean-floor basalt, and that the magnetic stripes parallel midocean ridges, it was natural to test the idea of Hess and examine the symmetry about midocean ridges. In less than a year, a publication appeared describing magnetic data across two ridge systems. Figure 2.1 portrays the symmetric magnetic anomaly patterns across a ridge system. The seafloor-spreading hypothesis was substantiated, and a stampede for further corroboration ensued.

★The Wadati–Benioff zone is commonly referred to in many publications as simply the Benioff zone, and *B-subduction* is the common expression that describes the subduction of oceanic crust. The omission of Wadati, even though his original paper preceded Benioff's is a consequence of Wadati's work being published in Japanese, and in this instance the ideas were not communicated to the scientific community at large.

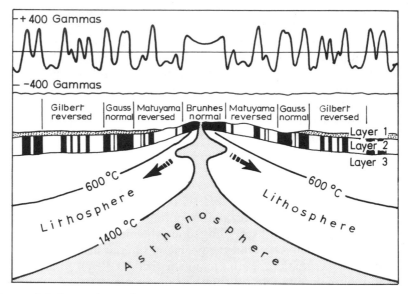

Figure 2.1 Formation of magnetic anomalies above the Curie temperature within Layer 2 of midocean-ridge ophiolites. (Modified from Cox and Hart (1987))

The Rosetta stone that broke the code to the kinematic puzzle of plate tectonics was provided by J. Tuzo Wilson in 1965. He showed how on the spherical Earth the relative motion of rigid plates results in three classes of boundaries (Figure 2.2). These classes include regions of underthrusting where plates are colliding (subduction zones), regions of extension where plates are spreading away from each other (midocean ridges), and sliding margins where plates move past each other (transform faults). The first two classes were the essential features of seafloor spreading; the third class integrated spreading and subduction into an organized system of plate motion. The sliding margins were termed transform faults because these fault surfaces transform (transfer) the plate motion from one ridge, trench, or other transform fault to another ridge, trench, or transform fault. The global network of interconnected ridges, subduction zones, and transform faults defines a family of large plates; the movements and resulting effects are what plate tectonics is all about.

The paradigm was now complete. Wegener in 1912 had provided the geometric framework, Wilson supplied the kinematic system, and Hess the fundamentals of the dynamic regime. What remained was an application of plate–tectonic principles to explain the vast array of geologic settings, the processes of mountain building, and the nature of continental growth.

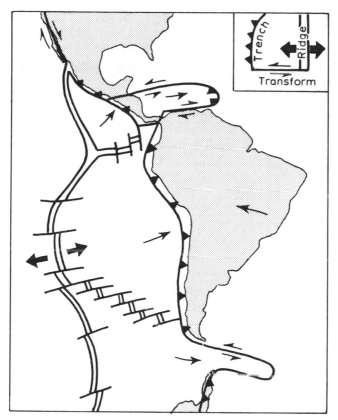

Figure 2.2 The three fundamental settings defining plate-boundary configurations: ridges where plates spread away from one another, trenches where one plate is consumed (subducted) beneath another, and transform faults where plates slide past each other. (Modified from Wilson (1965))

The nature of plates

The lithosphere consists of a mosaic of discrete relatively rigid plates. Plate thicknesses are controlled by the 1400° C isotherm, the temperature of partial melting that transforms the mantle into a quasiplastic medium. Thus, lithospheric plates are made up of both crust and upper mantle. Oceanic plates thicken away from the spreading ridges to an average thickness of about 60 km, and continental lithospheric plates average about 100 km in thickness. The oceanic portion of any given plate will expand or shrink in area depending on the distribution and rates of crustal generation and subduction associated with spreading ridges and subduction zones.

Since the early Archean, the total area of oceanic plates has gradually diminished as the area of continental plates has grown (see Chapter 3). The shapes and distribution of all plates are a consequence of continental tectonic effects and the orientation of subduction zones and midocean spreading ridges. In the current configuration, seven major plates (Africa, Antarctica, Eurasia, Indo-Australia, North America, South America, and Pacific) exceed an area of 10^7km^2. Six intermediate-sized plates (Arabia, Caribbean, Cocos, Nasca, Philippine, and Scotia) currently range in area from 10^6 to 10^7 km^2. Numerous plates exist that are smaller than 10^5 km^2. The importance of these small plates will be discussed below under the heading of microplate tectonics. The number and areal extents of individual plates are in a state of continual dynamic flux. The configuration of Pangea at 200 Ma represented a brief period when most (if not all) of the continental parts of the lithosphere were configured into one major continent, possibly even a single plate; the number of oceanic plates that existed at that time is not known as all but a few remnants have been subducted. A similar configuration of a supercontinental plate is inferred for the period of about 560 Ma when North America seems to have been located in the central region. Because continental fragments tend to cohere, at least momentarily, the long-term history of the shifting of plates has probably resulted in repetitive cycles of supercontinental agglomerations followed by intervals of continental dispersion and a consequent proliferation in the number of plates.

Plate boundaries display varying degrees of divergent, convergent, or translational movement. Plates are generally rigid and their relative motions can be visualized in terms of rotation about a fixed pole, the Euler pole of spherical geometry (Figure 2.3). Transform faults are always small circles about an Euler pole. Divergent and convergent boundaries are not small circles of rotation, and the orientation of these boundaries with respect to the small-circle trajectories may result in varying degrees of either convergence, divergence, or translational motion relative to the adjoining plate. The Aleutian Trench marks a plate boundary between the Pacific to the south and the North America plate to the north. Along the eastern part of the arc, the plate boundary is a convergent margin, but the obliquity of convergence increases toward the west until the trench is essentially a transform margin south of the Komandorsky Islands in the western part of the arc (Figure 2.3(b)). The pole of rotation for the Pacific plate, relative to the motion of the Indo-Australia plate, is situated on the Pacific plate just southeast of New Zealand. Near the Euler pole, plate motions swing about small circles of short radius. Along the relatively straight margin of New Zealand, motion on the incoming Pacific plate changes from orthogonal subduction in the north to strike slip in the south. A third example where plate motion changes along the strike of a margin is displayed where the Indo-Australia plate forms a subduction margin with

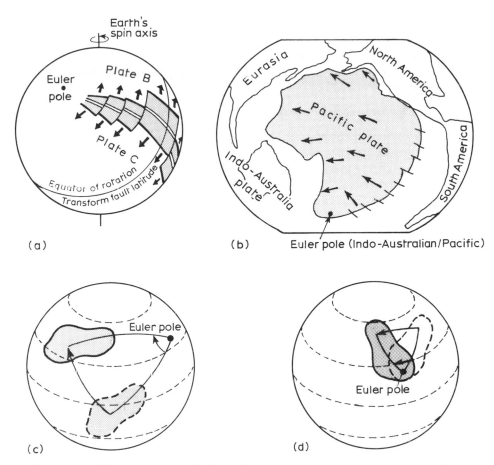

Figure 2.3 The movement of a rigid plate across the surface of a sphere is defined by the rotation about an Euler pole. The boundary conditions of a plate may range from spreading to sliding to consuming. Gradations among these three end-member configurations are based on the orientation of the rigid plates to the plate edges and to the motion vector defined by the Euler rotation; this situation is illustrated by the motion of the Pacific plate relative to the Indo-Australia plate in (b).

the Eurasia plate near Sumatra; but where the island chain curves to the north toward the Andaman Islands, the plate boundary becomes a transform fault, causing transpressional tectonics (Figure 2.3(b)).

The middle Paleozoic Caledonide orogenic system marks an ancient plate boundary. This ancient mountain system passes through Scotland and Scandinavia. The southwestern sector displays strike-slip or transform motion, presumably a small circle of rotation, but the northeastern part is typically characterized as a region of convergence and thrust faulting. Transform boundaries commonly show complex kinematic relations involving transpression and transtension, a characteristic that plays an important role in shaping continents and forming mountains and basins.

The principal driving forces moving plates

In Hess's seafloor-spreading model, there is a presumed viscous–drag force that couples the asthenosphere to the base of the lithosphere. Convection in the asthenosphere provides the dynamic force that moves the plates. A number of field observations, however, suggest that this force is not the dynamic factor of plate kinematics. For example, (a) in several places, such as south of Guadalcanal Island in the New Hebrides or west of southern Chile, active spreading ridges are being subducted, and the lack of any geophysical and geologic indications that spreading continues along the subducted portion of the ridge suggests that the plume of hot magma is extinguished. (b) The locus of spreading along segments of a ridge crest can jump laterally from tens to hundreds of kilometers; ridge spreading can also propagate longitudinally across a transform offset resulting in two overlapping parallel ridge segments. (c) In settings of closely spaced microplates, ridge axes and subduction zones are themselves necessarily closely spaced. These three observations make it unlikely that movement of mantle convection cells propels the more rigid lithospheric plates; otherwise, why would spreading stop after a ridge has entered a subduction zone? How can convection cells account for rapid shifts in the locus of spreading or longitudinal propagation of ridges? Finally, the close spacing of microplates seems to require either very small convection cells restricted to the upper mantle or deep convection cells having an unlikely height-to-width aspect ratio; both explanations seem unlikely.

Instead of plates actually being propelled by the convection cells, plates may simply slide under gravitational body forces. Movement of the lithosphere over the asthenosphere is along the interface defined by the 1400° C isotherm. This zone of partial melting reduces the friction between the two layers, although the gravitational body forces are opposed by *drag forces* along this zone. As the plate descends into the mantle the viscosity of the enclosing asthenosphere increases at the same time as the density of the downgoing lithosphere decreases, because of heating and mineral-phase changes. By about 700 km below the Earth's surface, the relative buoyancy forces are equalized, thereby arresting any further descent of the lithosphere deeper into the mantle.

In a dynamic model involving gravitational driving forces, the role of mantle convection is to maintain a mass balance within the asthenosphere and to provide the thermal bulges from which the plates descend. Because the cool crust and mantle of the oceanic lithosphere are more dense than the hot mantle of the asthenosphere, subduction progresses after a plate breaks and its leading edge sinks because it is less buoyant than the underlying asthenosphere. The gravitational sliding from the inflated ridge is commonly called *ridge push* and the sliding into the mantle at subduction zones is commonly referred to as *slab pull*, terms that may be somewhat misleading as the ridge is not pushing the plates, nor does the slab pull; both are simply the manifestation of gravitational attraction upon the entire body of a lithospheric plate (Figure 2.4). A ridge-push force is evinced in the Indian Ocean where the seafloor southeast of Sri Lanka is buckled into antiformal bulges. Ridge-push forces are also demonstrated by the America, Eurasia, and Africa plates, plates that are to a large degree surrounded by spreading ridges and almost nowhere along their perimeters are these plates composed of oceanic lithosphere descending into the mantle; nonetheless, these plates are moving, albeit slowly, and without slab pull. The importance of ridge push is manifest in the areas of intracontinental shortening, oftened refered to as *A-subduction* in honor of the Alpine geologist Ampferer. Intercontinental shortening is the principal mechanism responsible for the thickening of many of the world's major mountain systems, for example, Alps, Apennines, Andes, Himalaya, and Cordillera, but this ridge-push dynamic force remains poorly understood.

Where oceanic plates are involved, the vertical drop of oceanic lithosphere in subduction zones (generally 700 km) exceeds the vertical relief of the 1400° C isotherm (less than 100 km) away from ridge crests; thus, slab-pull forces are the more dominant of these two gravitational forces. This generalization is confirmed by the observations that plates with the greatest lengths of subduction zones move the fastest (Figure 2.5).

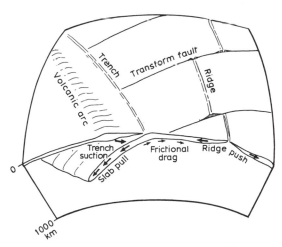

Figure 2.4 The four potential forces acting on a plate, the most effective force is slab pull followed by ridge push. See text for details. (Modified from Cox and Hart (1987))

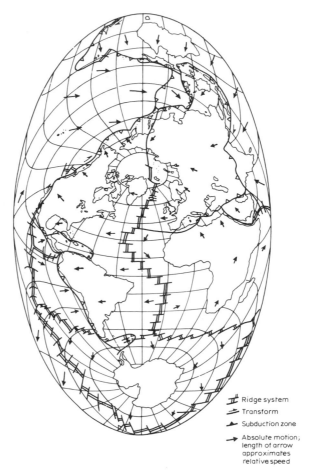

Ridge system
Transform
Subduction zone
Absolute motion;
length of arrow
approximates
relative speed

Figure 2.5 The absolute motion of plates, that is, the motion of plates relative to the core of the Earth. Note: the Antarctica plate is moving northward into the Pacific while the Indo-Australia and Pacific plates are moving more rapidly northward accommodating the lengthening of the circum-Antarctic ridge system; the Africa plate is rotating in a counterclockwise sense; and the America plate is sweeping westward while rotating counterclockwise in the North Atlantic to accommodate the slow northward advances of the Eurasian plate. (Data taken from Chase (1978))

Plate-tectonic settings

A brief review of various plate-tectonic settings is necessary at this point. Knowledge of the geometric and kinematic conditions in a relatively pristine state helps in understanding the complex relations that characterize most mountain systems of the world. Plate-tectonic settings are typically

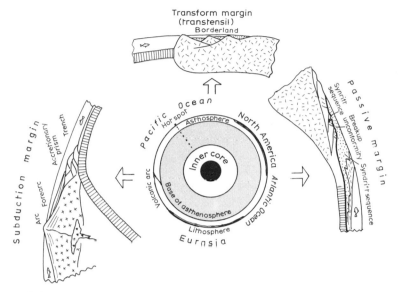

Figure 2.6 True-scale cross section of the Earth showing the relative thicknesses of the lithosphere, asthenosphere, and inner and outer core. Vertically exaggerated insets illustrate the principal features of subduction, rifted, and transform margins.

described in cross-sectional views. What may appear to be relatively minor differences along the strike, however, are commonly the features that provide explanations for particular phenomena, for example, transpression versus transtension, differences in the amount of sediment being subducted, changes in the angle of the downgoing oceanic plate in a B-subduction zone, the occurrence of an aseismic ridge, and numerous other situations.

Figure 2.6 shows a simple diagram of the basic crustal elements that result from plate-tectonic processes. Numerous other books and journal articles provide much greater detail about each of these features and the interested reader is encouraged to pursue this literature, but for the discussion at hand only a general understanding is required.

Midocean ridges

The midocean ridges are the sites where oceanic crust is formed. The elevation and breadth of the ridges depend directly on the rate of seafloor spreading. Slow-spreading ridges (total spreading at less than 3 cm/yr) are relatively narrow ridges with irregular tops and well developed axial valleys that only rise to about 3000 m below the sea surface. Fast-spreading ridges (greater than 6 cm/yr) are broad features with poorly developed axial valleys at bathymetric depths of about 2500 m. The mean depth of abyssal plains is approximately 5500 m; therefore, with a length of

56 000 km and an average height of 2500 to 3000 m, midocean ridges represent the largest mountain ranges on the planet Earth. Nonetheless, the ridges have hardly affected in any direct manner the tectonics of continents. Let me explain.

The midocean ridges stand high on the ocean floor as a result of thermal upwelling in the mantle. They do not, however, have deep crustal roots. Thus, when a ridge collides with a continent, such as offshore Vancouver Island, Canada, or southern Chile, relatively little, if any, structural disruption results. The ridge is subducted, and spreading ceases.

But what is the composition of the ocean crust? The basaltic extrusive component is commonly called MORB for *mid-ocean ridge basalt*. Basalt, and associated thin sedimentary cover, and an intrusive mafic and ultramafic layer represent the three divisions of an ophiolite (Figure 2.7). Ophiolitic sequences are found in many mountain ranges and older foldbelts and are commonly interpreted as former oceanic crust. But are ophiolites standard oceanic crust? The disussion about this may continue for decades, but the general consensus is that most, if not all, of the ophiolites enveloped in foldbelts are fragments not of standard oceanic crust but rather the crust from small short-lived ocean basins that formed offshore in close proximity to the locus of tectonic accretion. This assertion is supported by numerous geochemical analyses which indicate that the basalts of ophiolites are not MORB in character. Furthermore, the stratigraphic relations of many ophiolites indicate that the age of ophiolites are commonly only a little older than the age of their tectonic emplacement. This implies that the ophiolites must have formed in a region near the setting of their accretion, such as in a marginal basin, rather than in a remote mid-ocean spreading-ridge regime. The youthful aspect of marginal-basin ophiolites, relative to the age of tectonics in a contiguous convergence setting, may explain why

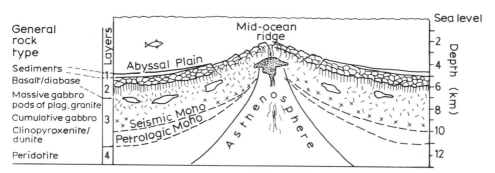

Figure 2.7 Schematic profile through a spreading ridge illustrating the vertical stratigraphy of an ophiolite.

they are preserved and MORB ophiolites are not–young crust would be hot and inflated, of low enough density that accretion rather than subduction is still possible.

Plateaus

Oceanic plateaus, sometimes referred to as aseismic ridges, are prominent features with crustal thickenesses commonly between 20 and 40 km thick. Hundreds of these features occur on all the ocean floors and range in size from small cone-shaped seamounts to enormous broad plateaus like Otong Java (Figure 2.8). The origin of most plateaus is currently a subject of great debate. The nature of the basement rock generally is not known. In a few places basement strata have been dredged from the flanks of a plateau, but more typically a thick sediment cover prevents direct access to the basement either by dredging or drilling. Seismic reflection and refraction data are ambiguous and allow a broad range of interpretations.

Most plateaus probably fall into one of four types: fragments of continents (for examples, Agulhus and Masquerine Plateaus), edifices built by ancient hotspots in a midplate region (for example Louisville and Sala y Gomez Ridges), thick accumulations of basalt formed by short lived thermal anomalies in a midplate region (for example Mid-Pacific Mountains and Galapagos Rise), and thickened crust that forms when a hotspot lies on a midocean ridge (for example Iceland today and possibly the Hess Rise). As passengers on the conveyer belt of ocean crust, they become accreted and produce tectonic effects during collisions along convergent margins.

Island arcs and subduction zones

Oceanic crust that descends into the mantle carries with it entrained seawater and probably small amounts of sediment. Hydrating a portion of the mantle causes mineralogic phase changes and some partial melting. This process produces the explosive volcanoes that lie above subduction zones (Figure 2.9). The ring of fire of the Pacific Ocean owes its existence to the subduction zones that rim the landward edges of the plates composing this ocean. The volcanoes occur either in an oceanic setting, such as the New Hebrides arc, or within a continental setting such as the Andean arc of South America. In either case, volcanic arcs are the principal domains for the generation of new continental crust. The growth rate of a volcanic arc, averaged over a normal 20 to 80 m.y. life span for an arc system, is generally 20 to 40 km^3/m.y. for each linear kilometer of arc. Worldwide, this equates to about 1 km^3/yr of new sialic material. What is

Figure 2.8 Map showing the general distribution of the larger oceanic plateaus, submerged regions of anomalously thick crust (15 to 40 km). Plateaus may be piles of basalt representing hotspots or some other form of midplate volcanism, remnant island arcs, or fragments of continents.

not certain, however, is whether or not this is a steady-state growth process. Volcanic arcs such as the Aleutian and Mariana systems seem to have had an initially rapid rate of growth that is followed by slower rates of growth. Chapter 3 provides an expanded discussion of this topic.

The structure of most volcanic-arc settings is unstable. Volcanic arcs often split, subduction zones may flip in polarity from one side of an arc to the other, and a subduction zone may abandon the arc, jumping to a new locality and a new orientation. Thus, even though subduction zones define the locus of generation for new sialic material, the arc massifs rarely survive intact; they move relative to continents, and continental growth takes place as a result of accretion tectonics involving arc–continent

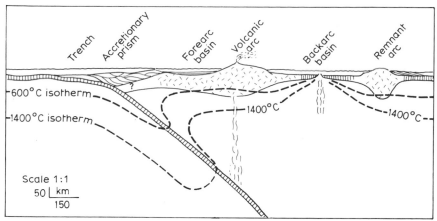

Figure 2.9 Diagram illustrating the principal features of an island arc with a backarc setting. Note the general configuration of the 600° C and 1400° C isotherms.

collisions; for example, Taiwan manifests the active accretion of a volcanic arc onto the east margin of Eurasia.

The architecture of volcanic-arc settings is principally controlled by the configuration of the subduction zone. In nearly all arc settings, the thermal reactions that produce the initial melting occur at a depth of about 100 km; therefore, the distance between the trench – the declivity in the seafloor marking initial descent of the crust – and the axis of volcanism varies as a function of the angle of the subducting plate. The considerable variation in these angles is demonstrated by plots of earthquake hypocenters defining the Wadati–Benioff zone (Figure 2.10). In areas of steep subduction the volcanic arcs are in close proximity to the trench – for example, the Northern Mariana arc in today's setting. Very low angles of subduction are also suggested; for example, during the middle Tertiary, 30–45 m.y. ago, arc volcanism along parts of western North America is thought to have lain as much as 2000 km eastward from the trench. This implies that the angle of the descending oceanic plate in the B–subduction was only about 3°. Where several subduction zones are in close proximity, subducting plates may be vertically stacked owing to variations in the angle of subduction; for example, along the east side of southern Japan the Pacific plate dips below the Philippine plate, yet both plates are subducting beneath Japan (Figure 2.11).

Other factors are probably related to the angle of subduction. These include: (1) the age of the subducting crust which reflects density differences – between 0 and 80 m.y. the aging of oceanic lithosphere corresponds to cooling and an increase in density; (2) the relative angle of motion between the two interacting plates; (3) the surface roughness of the incoming plate; and (4) the structural integrity of the descending plate – (a)

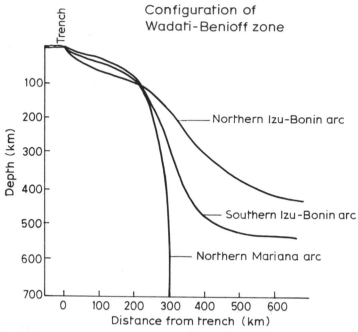

Figure 2.10 Locus of deep earthquakes beneath three volcanic-arc systems in the western Pacific illustrating variations in the Wadati–Benioff zone and the trajectory of subduction of the descending lithospheric plate.

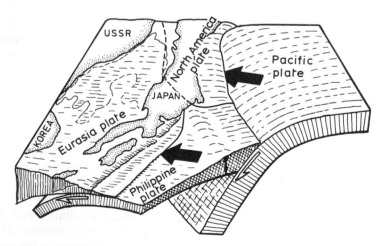

Figure 2.11 Block diagram displaying the vertical stacking of the Pacific and Philippine plates beneath Japan. Little wonder that Japan is exposed to a complex pattern of earthquakes.

the descending plate may break, allowing segments to move more rapidly into the mantle; (b) tears through the plate, oriented perpendicular to the trench, may cause the angle of subduction to change abruptly along strike, and (c) two different oceanic plates, for example, as mentioned above, the Philippine and Pacific plates south of Tokyo, Japan, can be subducted at different angles beneath a single point on a third plate. This creates a vertical stacking of multiple plates, each stimulating its own set of thermochemical reactions. It is little wonder that volcanologists are finding so much variation in the petrology, stratigraphy, and structure of volcanic terranes. These same sets of variations are important in the breakup and eventual accretion of arc fragments – volcanic terranes in orogenic foldbelts.

Between the volcanic centers and the trench lie two important structural settings. Directly trenchward from the arc lies a submerged part of the arc massif known as the forearc basin. Subsidence is a response to loading the crust by both the arc itself and subsequently by the vast volume of debris that is eroded from the emergent arc edifice. Farther trenchward is a more structurally disrupted region that is essentially the interface between the backbone of the arc and the descending oceanic plate, commonly referred to as the accretionary prism (Figure 2.9). The precise interface between the upper and lower plates has been imaged seismically in a variety of places and penetrated with the drill in the Lesser Antilles arc system. This information has revealed a decollement 20 to 40 m thick where the hydrostatic pressure approaches that of the lithostatic pressure. The level of the detachment horizon tends to migrate downward toward the oceanic crust, commensurate with dewatering of the clays, as the descending plate advances into the subduction zone. The consequence of this downward migration and dewatering is to structurally transfer material that is too light to sink into the mantle from the lower oceanic plate to the volcanic-arc upper plate. The structural styles affecting this transfer, or accretion, are quite variable owing to all the parameters of the subduction process mentioned above. In general, however, two effects result: accretion at the toe of the accretionary prism; and *underplating*, sometimes called sub-cretion, beneath an interior portion of the accretionary prism.

An important observation made during tectonic investigations is the realization that crustal material deformed in compressive regimes will tend toward a predictable geometric configuration that depends on the strength of the material, fluid pressure, and internal friction. Simplistically stated, fold–and–thrust belts and accretionary wedges of subduction zones behave in a manner analogous to that of a wedge of snow or soil in front of a moving bulldozer (Figure 2.12); these materials deform internally until arriving at a shape that is characterized by a *critical taper*, whereupon the mass slides without further internal deformation along a rigid basal decollement. Any material added or accreted to the wedge causes further

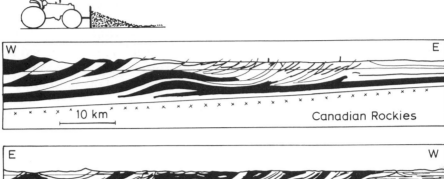

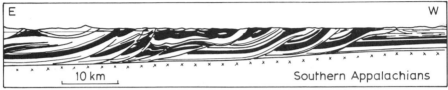

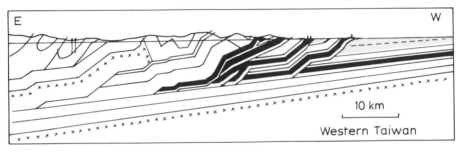

Figure 2.12 Cross sections of several compressional fold-and-thrust belts indicating the wedge-shaped profile – critical taper – and comparing it to the deformation of a moving bulldozer. (Figure modified from Davis, Suppe, and Dahlen (1983), profiles from Bally, Gordy, and Stewart (1986) (Canadian Rockies), Roeder, Gilbert, and Witherspoon (1978) (southern Appalachians), and Suppe (1980) (western Taiwan))

internal deformation until the geometry of the critical taper is again achieved. The dimensions of the critical taper depend on the fundamental properties of internal and basal friction, internal and basal fluid pressure ratios, and the densities of the wedge material and that of water.

Because crustal materials move along a rigid base only after the critical taper is formed, geologists can now explain why extensional faulting is commonly observed in a compressional regime. Imagine a subduction zone where the accretionary prism is stabilized in the form of a critical taper. If the inclination of the subduction zone should steepen, the accretionary wedge would no longer have the critical geometry. The new wedge would be subcritical, that is too thin. Thrust faulting and the accretion of new material are needed to thicken the wedge, creating the requisite taper (Figure 2.13). In contrast, if the descending slab of the subduction zone should flatten, the wedge would be supercritical, that is

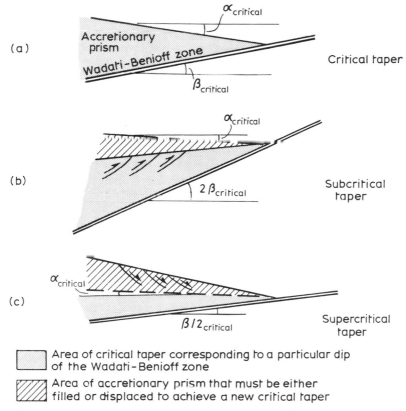

Figure 2.13 Illustration of the stages of a fold-and-thrust belt. (a) The profile has achieved the critical taper, and therefore the mass will slide along the rigid base without further internal deformation. (b) A profile with a subcritical taper as a consequence of the steepening of the rigid base; in this case the material will shorten by thrust faulting in order to elevate the topographic plane and restore the critical taper. (c) A crustal profile with a supercritical taper owing to the shallowing of the inclination of the rigid base. In these instances the mass will fail by extensional faulting in order to achieve the critical-taper configuration. In each case the angle of the rigid base corresponds to a unique angle for the topographic profile defined by the relationship of $\alpha_{\text{critical}} + R\beta = F$, where R and F are functions of friction, fluid pressure, and material densities. See Davis, Suppe, and Dahlen (1983) for an elaboration of mathematical relationships. (Figure taken from Lallemand (1987))

too thick, necessitating internal extensional deformation to thin and lengthen the wedge into a new critical taper. Many ways can be envisaged to cause steepening or shallowing of the basal decollement of an accretionary wedge or fold-and-thrust belt, for example, a sudden change in the age of the oceanic crust, a change in the rate of tectonic closure, the arrival of an oceanic plateau to the subduction zone, the decollement ramping upward to a higher level, thicker continental crust arriving at the point of thrusting, and so forth. The newly popularized concept of a critical taper is now being applied to the tectonic history of most mountain systems. In Chapter 6, where the southern Andes are discussed, the notion of a critical taper will be developed to explain the apparent absence of accretion along the west margin of Peru and Chile.

Not all volcanic arcs have well developed accretionary prisms. For instance, along the Peru–Chile and Middle America Trenches, old continental basement rock occurs within 10 to 12 km of the trench axis. Additionally, paleontologic and sedimentologic analyses have demonstrated intervals of massive subsidence within the outer forearc regions of some convergent margins. This seems to require the removal of material, referred to as *subduction erosion*, from the front and from beneath the upper plate in order to account for the collapse. Has the missing portion been transferred either laterally along the front of the arc, or downward into a deeper region of the upper plate? But if the lightweight crustal material is subducted into the mantle, one must ask how? It seems improbable that the adhesive forces needed to bond slivers of a truncated continental margin to the subducting oceanic lithosphere would be strong enough to overcome the buoyancy forces exerted on the low-density continental crust as it descends into the high-density mantle. Even more puzzling is the assertion by some geologists that sediment – essentially particulate continental crust – is subducted into the mantle. This subject will be discussed in greater detail in Chapter 3, as a recurring theme in this book is the staying power of continental crust. Once formed, continental crust generally remains in the crustal environment – no matter what form of transformation may occur – simply because of its buoyancy.

Some of the confusion involving the processes and occurrences of subduction erosion may stem from a misunderstanding of what is meant by subduction. Material passing beneath the deformation front of a subduction may constitute 'subduction', but this does not mean that all the material continues the descent into the mantle. The low–density material is most likely tectonically transfered to the underside of the upper plate in a middle-to-lower crustal horizon.

The principal means of thickening continental crust is by intracontinental thrusting, the process referred to as A-subduction. This form of subduction is contrasted with B-subduction, where an oceanic plate plunges beneath a continent. A-subduction is an important phenomenon, and examples will be discussed in Chapter 6. One must not conclude, however, that this form of subduction implies that the continental crust is recycled into the mantle. As will be shown, the continental crust more likely 'doubles' in thickness as a result of intracrustal thick-skinned thrusting along decollement surfaces of the lower crust. Indeed, this is the principal phenomenon responsible for the formation of thick crustal roots. beneath orogenic mountain belts.

Backarc basins and marginal seas

Behind★ some volcanic arcs are oceanic basins, and some of them have formed after the initiation of volcanism. In these regions the arc splits and seafloor spreading begins behind the locus of active volcanism. Backarc basins are common in the southwest Pacific Ocean but are much rarer elsewhere. The Lau basin is a backarc basin to the Tonga arc and the Shikoku basin is a backarc basin to the northern Mariana arc. The dynamics of how volcanic arcs split to form backarc basins are not well understood. A variety of geometric and kinematic parameters seem to be attendant upon the occurrences of backarc basins; these parameters include: subduction of old ocean crust, which favors, a steep Wadati–Benioff zone; the absolute motion of the upper and lower plates are not opposed to each other (Figure 2.5); and an absence of ridge-push forces in the upper plate. All three parameters reflect conditions that minimize drag forces. Therefore, because of the apparent absence of forces that could hold the crust together, and because the crust in the region of the arc presumably is softened by thermal processes, the crust fails and seafloor spreading commences.

The Japan and South China Seas are types of backarc basins, but they probably owe their initial existence to regional transtensional slip between the Eurasia and Philippine plates (large pull-apart basins) rather than to the splitting of an arc. The Bering Sea is in a backarc setting also, but that basin is an example of normal midocean crust that has become tectonically

---

★Convention of usage says that volcanic arcs *face* the trench; therefore, in front of the arc, that is toward the trench, is the forearc region, whereas behind the arc is the backarc region (sometimes called reararc). This nomenclature creates a geometric puzzle. The facing direction of sedimentary layers is in the direction toward stratigraphically younger sediments. Prior to any tectonic disruption the facing direction is always up, skyward; but, because of structural disruption, in many accretionary prisms individual beds face toward the arc while the age of sediment packages within the prism become younger toward the trench. If you can solve this puzzle, you understand the basic geometry of offscraped sediments that form at the toe of an accretionary prism. See Chapter 6 for a solution.

trapped in a backarc setting. A variety of kinematic explanations are possible in a trapping scenario. In the simplest mode, a subduction zone jumps oceanward preserving the intervening segment of ocean crust between the abandoned and the newly established zones of subduction. When the subduction zone jumped southward from the Anadyr–Koryak Mountains of eastern Russia to the region of the present Aleutian–Komandorsky Islands at approximately 50 Ma, a segment of 140 to 120 m.y. old crust, the present basement of the Bering Sea, was trapped behind the newly formed volcanic arc (Figure 2.14). The Caribbean and Scotia Seas are another variety of backarc basin where old oceanic crust is trapped behind a more youthful volcanic arc. In these cases the older oceanic crust represents prongs of oceanic crust torn from the Pacific Plate and moved relatively eastward. The westward movement of three continental masses (North and South America and Antarctica) left two gaps between them (the Caribbean and Scotia Seas).

Some uncertainty remains regarding the origin of many other backarc basins. The Caspian and Black Seas are thought to be possibly trapped oceanic crust. Geologists continue to debate whether or not the Banda, Celebes, and Sulu Seas are segmented portions of a trapped piece of the old Indian plate or basins reflecting transtensional tectonics within the complex network of volcanic arcs that make up the Indonesian and

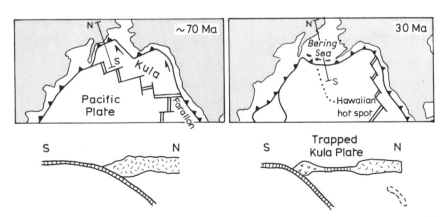

Figure 2.14 Trapping mechanism for the oceanic crust of the Bering Sea. The southward jump in subduction took place about 50 Ma ago, isolating a piece of the Kula plate in a backarc setting north of the newly formed Aleutian volcanic arc. The accretion of Shikoku Ridge and Umnak Plateau, as well as accretionary tectonics in southeastern Alaska, may have temporarily 'clogged' the northward subduction along the southeast margin of Siberia and Alaska, forcing the southward jump in the location of subduction.

Philippine archipelagos. An important clue that will help answer these questions involves determining the age of the basin crust, either by direct sampling or by magnetostratigraphy.

On a more theoretical basis, there is another dynamic regime that could result in the fracturing of the upper plate of a subduction zone. Understanding this basis requires introducing an additional type of hypothetical plate-driving force, with the typically ambiguous appellation *trench suction*. In situations where the upper plate has a component of absolute motion toward the trench, the position of the trench must migrate oceanward, relative to a fixed frame of reference in the mantle below the landward plate. This migration of the trench is called *trench rollback*. Trench rollback occurs along the west margin of North and South America except for the short gaps represented by the Caribbean and Scotia Seas. The condition of trench rollback requires the descending lithosphere to retreat broadside through the asthenosphere. The retreat causes a mass excess on the oceanward side of the descending lithosphere and a mass deficiency on the arc side. To compensate for this, convective flow in the asthenosphere is directed toward the zone of mass deficiency, and this produces a suction force upon the upper plate. Where this is happening along the Pacific side of the North and South American plates, trench rollback and the associated suction forces augment ridge-push forces resulting from Mid-Atlantic Ridge spreading. If the trenchward advancement of the upper plate is impeded, at one or more points – possibly by the collision of oceanic plateaus – trench suction forces could cause the upper plate to fail in the intervening area where the advancement is not obstructed. The basins of the Okinawa trough are possibly representative of such a situation. In this instance, the collision of the Luzon arc and the Palau–Kyushu Ridge have pinned two points of the Eurasian margin, while the upper plate between these two points tears apart owing to trench suction (Figure 2.15).

Plate tectonic reconstructions

The continental drift reconstructions of Wegener and his colleagues were constrained principally by interpretations of the outlines of continents. The dictates of the seafloor-spreading model and magnetostratigraphy provide constraints for reconstructing the locations of plates in specific time frames during the past 200 m.y., since the breakup of Pangea. These data are not useful in constraining reconstructions older than 200 Ma, the age of the oldest preserved oceanic crust, yet there is no reason to think that seafloor spreading and continental drift are restricted to recent geologic history.

Studies of ancient foldbelts indicate repeated epiosdes of orogenic activity, which most likely reflect the breaking up and colliding of continents. The tectonics and timing of such events were known in the late

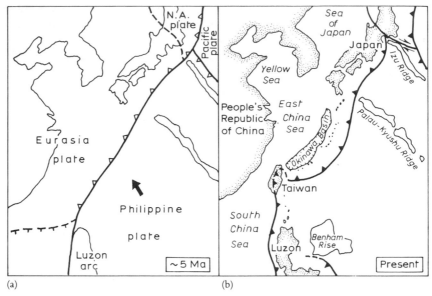

Figure 2.15 Collision tectonics and the formation of a basin in the upper plate. (a) The arrival of three prominent ridges (Izu, Palau–Kyushu, and the Luzon arc) to the subduction zone along the east margin of the Eurasia plate. The absolute motion of the Eurasia plate is only slightly oblique to the motion of the Philippine plate; thus, trench rollback and trench suction forces would be small. The eastward expulsion of fragments of Asia resulting from the northward indentation of India into Asia, however, may stimulate further this rollback phenomenon (see Figure 4.16). (b) The opening of the Okinawa basin involves 'pinning' the subduction at two points; trench suction forces preferentially move the upper plate in the region between colliding ridges and thereby effect an extensional tear (Viallon, Huchon, and Barrier, 1986).

1800s for the Appalachian–Caledonian system. North American and European geologists were familiar with the geology in this system, and the Atlantic Ocean seemed to have opened and closed several times during the Paleozoic to account for the discrete episodes of metamorphism and tectonism. Many fossil localities within the region of folded strata contained assemblages that were endemic to the continent on the opposite side of the Atlantic. This suggests that continental fragments were left behind during renewed ocean opening followed by continental drift. The opening and closing★ of oceans became known as the Wilson cycle in honor of J. Tuzo Wilson.

★ The statement that an ocean closes is in some instances misleading. For example, from the late Paleozoic to the present numerous accretionary events have incorporated island arcs and

Except for small silvers of ophiolitic material kneaded into the Appalachian–Caledonian foldbelt, no voluminous record of ancient oceanic crust has been preserved. Nonetheless, some of these oceans have been named. Iapetus, father of Atlantis in Greek mythology, is the name given to the complex ocean system that started to form in the late Precambrian and experienced as many as three 'closing' episodes during the Paleozoic – corresponding to orogenic pulses known as the Taconic, Acadian, and Alleghenian orogenies of the Appalachian foldbelt of eastern North America and the Caledonian and Hercynian orogenies for the foldbelts of western Europe. Other lost oceans include Panthalassa, the ocean that surrounded Pangea prior to fragmentation, and the Paleo- and Neo-Tethys Sea, equatorial passages that separated Gondwana from Eurasia. Other episodes involving the accretion of Gondwana fragments must have preceded the 43 Ma closure of the Neo-Tethys – when India slammed into Asia–along the south margin of Eurasia. The so-called Paleo-Tethys closed during the middle Mesozoic; these welded fragments of the Cimmeria (now represented by continental terranes stretching from Indochina to Turkey) onto the south margin of Eurasia. Australia is destined to be the next bit chunk of Gondwana to close against the Asian margin, yet an oceanic realm will persist behind the trailing margin.

What were the shapes of these oceans and what global routes did the continents follow as they drifted along their ancient paths? For periods prior to 200 Ma ago matching continental margins, correlating crustal provinces and stratal horizons, and reconstructing zoogeographic provinces do not, by themselves, provide enough information to constrain the locations of wandering continents accurately. The application of paleomagnetic data, however, when added to these other parameters, greatly improves our ability to reconstruct the travel paths of continents.

Most of the Earth's magnetic field can be modeled as a geocentric dipole, essentially a bar magnet oriented roughly coincident with the Earth's spin axis. It has already been mentioned that the north and south poles of this dipole flip back and forth, providing the basis for magnetostratigraphy; a

(Footnote continued.)

continental fragments along the south margin of Asia. In each instance, large tracts of oceanic crust have been consumed; the loss of the Paleo- and Neo-Tethys Sea is discussed in the text. Nonetheless, after each of these episodes of accretion, the continental margin remained facing an open ocean to the south. Similary, the west margin of North America faced an open ocean for the past 600 m.y., yet oceanic plates such as the Kula and Farallon have closed along its margin. In these instances 'closure' means the subduction of an oceanic plate. But when Africa, South America, Eurasia, and North America joined to form Pangea in the late Paleozoic, not only was the oceanic crust wholly subducted, the intervening oceanic setting was also lost.

second important attribute is the orientation of the magnetic force field in relation to the Earth's surface. Imagine a bar magnet surrounded by minute iron filings which align themselves along the force field (Figure 2.16). The intersection of the force lines with the surface of the Earth varies as a function of latitude. At the equator the field is parallel to the Earth's surface, and the *magnetic inclinations* steepen systematically toward the magnetic poles, where the inclination is vertical to the Earth's surface. If the sense of magnetic polarity is not known for the time when a rock freezes the orientation of a magnetic signal, interpreting either a northern or a southern hemisphere position is possible. Other forms of geologic data may help to resolve this ambiguity.

Rocks which preserve the ancient magnetic inclinations through a thermal remanent magnetization record the latitudinal position only at the time when the magnetic carrier passes below its *blocking temperature*, that is, the temperature that allows the magnetic particles to freeze in partial alignment with the Earth's magnetic force field. For many magnetic minerals this is in the range of 500° to 600° C, roughly equivalent to blocking temperatures for argon in biotite and hornblende, the temperature where argon begins to be retained in the crystal lattice. Consequently, K/Ar ages provide the time of acquisition for some thermoremanent magnetism. Typical seafloor basalt, however, contains magnetite high in titanium, and this has a lower blocking temperature of about 200° C.

Other forms of remanent magnetization include detrital and chemical remanent magnetization. Detrital remanence results from the partial alignment of magnetic detrital particles with the magnetic force fields during epiosdes of deposition. An alignment of detrital magnetic particles may also reflect the orientation of flow of the transporting fluid; however, studies have also shown that the so-called detrital remanence may be post-depositional, in that prior to the dewatering and compaction of sediment, the tiny magnetic particles are free to move about in the interstices of the detrital framework. Chemical remanence is often associated with the post-depositional alteration of a rock, such as weathering, low-grade metamorphism or the introduction of fluids and the precipitation of cements.

Besides inclination data that relate to paleolatitude, remanent magnetism may also preserve a remanent compass orientation (*magnetic declination*); a bearing, which if it diverges significantly from either the north or south poles, indicates a tectonic rotation. Along the west margin of North America, paleomagnetic data commonly show declinations oriented east − 'to the right' − of magnetic pole positions for coeval strata from the craton that lies east of the margin. Thus, a clockwise sense of rotation is implied, consistent with the tectonics of a right–lateral strike–slip regime. This rotation is a consequence of the relative motion of the North America plate and the oceanic plates to the west during the past 110 m.y. This sense

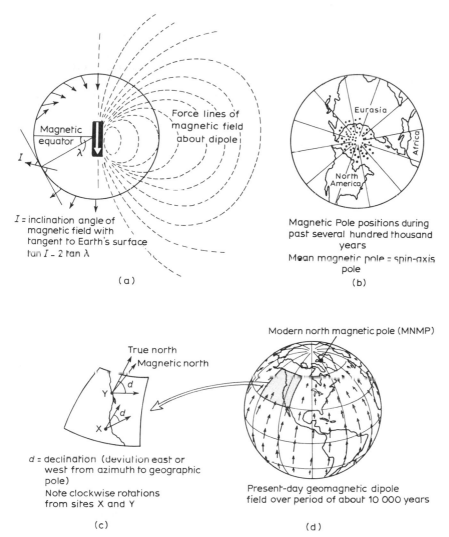

Figure 2.16 Cross sections of the Earth. (a) The orientation of the magnetic field with a magnetic dipole pole coincident with the Earth's spin axis. The 'axial geocentric dipole model' is consistent with mean magnetic pole positions. (b) Approximate recordings of the magnetic pole positions during the past 10 000 years. Because the dipole precesses about the spin axis, at any instant in time the magnetic poles will lie to the side of the spin axis. For example, the modern north magnetic pole lies near northwest Greenland. (c) Paleomagnetic deflections away from the paleomagnetic pole position; this may indicate a tectonic disturbance. This amount of deflection, or 'declination', can be a direct measurement of the amount of crustal rotation, for example, along western North America. (d) Clockwise rotations associated with right-slip faulting.

of rotation can be visualized by placing a vertically oriented pencil between your hands, as though you are praying, and sliding the left hand away from your body. The pencil should rotate in a clockwise sense under this right-lateral shear. Sometimes geologists call this 'ball-bearing' tectonics.

Many pitfalls may attend paleomagnetic studies, such as multiple episodes of magnetic overprinting, weathering and degradation of the remanent magnetic signal, recording short–term precessional events of the dipole (so–called secular variations), uncertainties regarding the time of magnetic acquisition, uncertainty about the orientation of the strata at the time of magnetic acquisition, as well as possible errors arising from instrumental imprecision. Tests and procedures have been developed to guard against these problems, but data constantly need to be re-evaluated. Field geologists, who may be skeptical about black-box solutions – especially solutions that imply tectonic movements that are not otherwise immediately evident in 'their' rocks – generally say that before one dares accept the results of a 'paleomagician' at least two independent laboratories must obtain consistent results. And even with these confirming results, some geologists refuse to believe in the efficacy of paleomagnetism for tectonic, kinematic studies.

One of the early objectives of these kinds of paleomagnetic studies was to track the positions of all the major continents relative to the magnetic poles, inferred to be always coincident with the poles of the Earth's spin axis. These data can be displayed in a reference frame where the positions of the continents are held fixed and the magnetic pole(s) are assumed to have moved; the *apparent polar wander* (APW) path for a given continent is a reconstruction of paleomagnetic poles of many different ages relative to a spatially fixed continent. The travel history of continents can be analyzed by plotting their APW paths (Figure 2.17). Adjoining continents that have parallel APW paths must have remained fixed relative to one another for the duration recorded by their respective coincident APW paths. The APW paths of two continents experiencing a Wilson cycle will show parallel tracks that diverge at the time of breakup and converge to a second interval of parallel tracks following ocean closure. Commonly the ATW paths of continents are much more complex, recording the convolutions typical of plate motions.

In one circumstance, involving large continents, APW tracks enable Earth scientists to reconstruct uniquely the absolute rate of motion, the paleolatitude, and the paleolongitude. In order to make such a reconstruction, the APW tracks must lie along a smooth line, describing a small circle of rotation. In these occurrences, the direction of motion had to be constant. By calculating a paleomagnetic pole of rotation (Euler pole) for this small circle, and by knowing the age for the various points along the APW track, the absolute plate velocity is calculable. Recent testing of this approach has

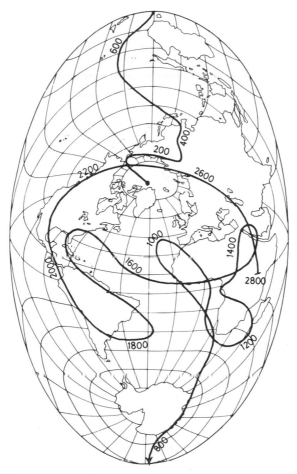

Figure 2.17 This spaghetti-like curve is the generalized apparent polar wander (APW) path for the Archean cratonal part of the North American continent as well as all crustal fragments that have accreted to it since 2800 m.y. ago. To comprehend the drifting of continents, one must know that in reality the magnetic pole has presumably remained fixed at the spin axis, and it is the North American continent that has wandered. The distance between the APW path and a site on the continent gives the latitude of the site for each time interval. No uniform relationship exists between the length of the path and the duration of the record. Except for straight-line segments along this path, the longitudinal position of the continent is not constrained. Therefore, the spatial trajectory of the North American continent is unlikely to be similar in shape to the APW path. East–west excursions between data points cannot be distinguished from stationary intervals along the APW (which are not discernible with only 200 m.y. tick marks as shown here). (Figure modified from Tarling (1981) and McElhinny (1973))

yielded some surprising results: whereas large continents are now moving at absolute rates less than 2 cm/yr, at intervals before 140 m.y. ago, both the Eurasia and North America plates were moving at a rate of 6 to 10 cm/yr. Possibly each of these continental plates was along the trailing margin of a large oceanic plate that was subducting and therefore pulling the continents at a rate more rapid than is realized by the ridge-push forces operating today on all large continental plates. India provides a parallel situation with respect to its kinematic history since breaking away from Gondwanaland in the middle part of the Cretaceous.

Except for the just-mentioned special instance involving small-circle APW tracks associated with large continents, paleomagnetic data do not constrain paleolongitudinal positions. Only the latitude, the distance from the magnetic pole, can be determined from inclination data. For any given inclination the sample may lie anywhere along the 360° of longitude defined by these data. Thus, even with a good set of paleomagnetic data, inherent uncertainties remain. Combining geologic and paleomagnetic data greatly limits the paleogeographic options. Reconstructions depicting the breakup of Gondwanaland are rather precisely constrained by the pattern of magnetic anomaly isochrons mapped throughout the southern oceans. Cretaceous and younger reconstructions of the major plates are constrained by ocean magnetic-lineation data whereas pre-Jurassic reconstructions must incorporate paleomagnetic-inclination data, APW data, and geologic considerations, such as sedimentologic criteria consistent with equatorial or high-latitude depositional settings. Global paleogeographic reconstructions are also available as computer software programs allowing the user to incorporate the latest data and make modifications that might better satisfy a particular geologic relationship. An encouraging consensus supports the general makeup of these reconstructions. Certainly, many of the major orogenic episodes for the past 600 m.y. can be placed in the context of some form of a continental collision. Many of the imagined offshore landmasses of the geosynclinal period can now be explained in terms of the wanderings of continents.

Microplates

Even though most of the Earth's surface is covered by seven large plates, areas such as those exemplifed by the Indonesian and Philippine archipelagos contain many small plates; compared with plates like the Pacific, some of these are microsized plates. Microplates interact amongst themselves in a complex fashion both in space and time. Because of the high incidence of collisions throughout the network of microplates, the distribution of subduction zones is unstable and changeable: arcs commonly rotate, relative plate motions change rapidly along the strike of a plate boundary, and the direction of subduction can flip in polarity – alternating from one side of an arc to the other side. The high density of plate boundaries creates a complicated stress field, and one cannot insist on the notion of the rigidity of plates. The tectonics of microplates is exciting, and many of the mountain systems of the world bear characteristics that can be ascribed to an ancient microplate tectonic setting. This subject therefore is expanded in subsequent chapters.

2.3 CONCLUSION

Our understanding of crustal processes has advanced greatly since the heyday of geosynclines. Paleogeographic reconstructions are more rigidly constrained both geometrically and kinematically than in the early continental-drift scenarios, and plate tectonics provides a dynamic model enabling geologists to predict ancient settings even though most of the record has been lost. Paleomagnetic studies are particularly helpful in constraining both spatial and chronologic aspects of crustal plates. Many studies have demonstrated that the geologic history of continents during the past 200 m.y. is better understood with a knowledge of the tectonic history of the adjacent ocean crust.

Nonetheless, many problems still exist. Geologic mapping in places like Alaska, Japan, or Turkey demonstrates that simplistic applications of the plate-tectonic model do not explain the distribution of rocks in modern orogenic foldbelts. Not all plates or continental fragments have well established APW paths. The role of microplate tectonics in the grand scheme of global motions of the larger continental plates is often lost in continental-scale paleogeographic reconstructions. Marginal regions of many continents contain strata that have paleomagnetic signatures which do not correspond to the APW traces of the cratonal portion. Global-scale reconstructions do not show the effects of island-arc or plateau accretion, or how continental margins respond to collision and rifting events. These are the objectives of terrane analysis, and these details are necessary to understand the processes of mountain building and the growth of continents.

History of continental growth

3.1 CRUSTAL RECYCLING

Are continents growing? Geologists offer extreme points of view in answer to this question. One school proposes that the subduction process recycles at least $1.5 \, km^3/yr$ of crustal rock into the mantle. Because this is approximately equal to the production volume of igneous material in island arcs and above hotspots, it would suggest that the volume of continental-type rock is in a steady state. An opposing viewpoint holds that little or no material is recycled into the mantle, and therefore the volume of continental-type rock must be growing. The heart of the issue is understanding the nature of lithologic recycling. Within the crustal domain, lithologic recycling involves both multiple cycles of resedimentation and anatectic processes where sediment is melted and reintroduced as an igneous product. But does exchange take place between the crust and mantle, or is recycling of rocks principally an intracrustal phenomenon? In trying to answer this question, we must make a distinction between oceanic and continental crust.

Oceanic-crust recycling

The circumstance of oceanic crust provides an almost perfect example of steady-state recycling between the crust and the mantle. Figure 3.1 uses both a histogram and a cumulative curve to portray the age distribution of oceanic crust. The mean age of the oceanic crust is 55 Ma. Even though oscillations in sea level during the past 550 m.y. indicate probable fluctuations in short-term spreading rates (Figure 3.2), the 55 Ma average age indicates a mean rate of crustal spreading since 200 Ma of 5.03 cm/yr (full rate). This is the velocity of one plate relative to its pair moving in the opposite direction. Crust that is born along ridges undergoes an aging process *en route* to its demise in subduction zones, where it is recycled back into the mantle. Little of this igneous crustal material falls out of the recycling loop. The flux of ocean crust is essentially a closed system; what is added at one place is subtracted from another. With a total volume of

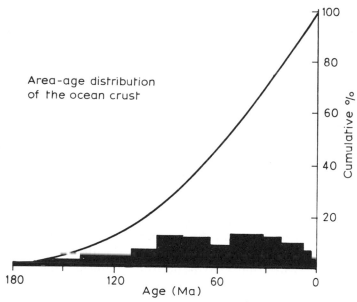

Figure 3.1 Distribution of the $309 \times 10^6 \, \text{km}^2$ of oceanic crust at the Earth's surface as a function of age. The age of oceanic crust ranges from zero to approximately 180 Ma with a mean of 55 Ma, indicating that oceanic crust is recycled every 110 Ma, although the rate of recycling may have been greater in the past. The histogram of ages is not symmetric about the mean because (a) ocean spreading ridges are not everywhere in a midocean setting, (b) the rate of spreading varies along the length of the ridge system, and (c) the rate of spreading may be asymmetric.

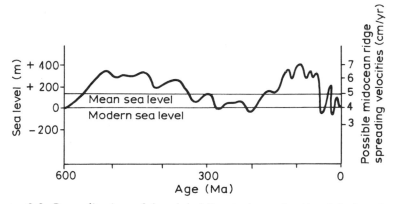

Figure 3.2 Generalization of the global fluctuations of sea level during the past 600 m.y. The 400 m sweep of eustasy reflects principally changes in the mean spreading rates along ocean ridge spreading axes coupled with effects of glaciation in the early and late Paleozoic and the late Cenozoic.

oceanic crust of $1.8 \times 10^9 \, km^3$ (6 km thick and an area of 300 million km^2), and with the population characteristics portrayed in Figure 3.1, approximately $17 \, km^3/yr$ must be recycled in order to maintain the steady-state configuration.

Oceanic sediment

The mass of oceanic sediment can also be analyzed in the context of recycling dynamics. The age distribution of ocean sediment is not as well constrained as that for igneous oceanic crust (Figure 3.3); nonetheless, a cumulative distribution curve can be fitted to the data, which, assuming a steady state, requires a recycling volume of about $1.5–2.0 \, km^3/yr$. An influx of sedimentary rock to the seafloor – underlain by ocean crust – of $1.7 \, km^3/yr$ has been determined by dividing an estimate of the total volume of sediment lying on the oceanic crust ($1.14 \times 10^8 \, km^3$, when converted to a density of $2.7 \, g \, cm^{-3}$) by 55 Ma, the average age of the ocean crust. But what happens to the deep-sea sediment as the ocean crust is subducted? Is the recycling system open or closed? Is the sediment also subducted, or does it remain in the crustal domain by being scraped off onto the upper plate? Before addressing this subject, we must first explore the possible recycling scenarios of continental crust.

Continental-crust recycling

The current volume of continental crust is approximately $7.6 \times 10^9 \, km^3$. Figure 3.4 plots several data sets that compare age of basement as a function of cumulative volumes of continental crust.

The different slopes for the regression curves of Figure 3.4 reflect dating techniques of different isotopic systems. Systems such as K/Ar have relatively low blocking temperatures, that is the mineral lattices become open systems when heated to $200–600° \, C$ – the exact temperature being dependent on the particular mineral. Heating a mineral above its blocking temperature has the effect of resetting to zero its radiogenic clock. Mineral or whole-rock dates involving Rb/Sr and U/Pb have progressively higher blocking temperatures and therefore tend to express an older age distribution. Of all the dating techniques, Sm/Nd holds the greatest promise of providing a chronicle of when continental basement rocks first crystallized. The similar geochemical properties of both the parent Sm and the daughter Nd mean that they behave nearly identically regardless of the crustal environment. The two elements do not fractionate upon heating or weathering. Thus, the Sm/Nd whole-rock age of many samples, regardless of geologic history – weathering and metamorphism – will approximate the time when the rock first fractionated out of the mantle. This is a relatively new dating technique. Enough information is known, however,

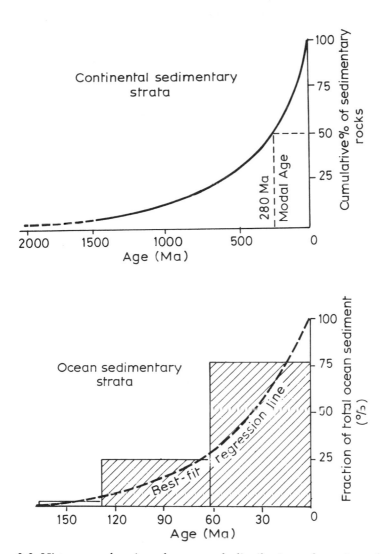

Figure 3.3 Histogram showing the general distribution of continental and oceanic sediment. Note the change in scales reflecting the younger age of oceanic crust compared with that for continents. The average age distributions fit regression analyses suggesting that the sediments are recycled. The 280 Ma average age for the continental sediments indicates a recycling rate that is much more rapid than the growth rate of the basement. In the case of the oceanic domain, the recycling rate seems to be slightly more rapid than the recycling rate of the basement. The larger percentage of younger sediment compared with the age distribution of oceanic crust is probably an effect of increased sedimentation rates as a result of (a) the extraordinary continental uplift with increased rates of erosion in southern Asia after the accretion 43 m.y. ago of India, and (b) glaciation during the late Cenozoic. (Modified from Veizer and Jansen (1979, 1984))

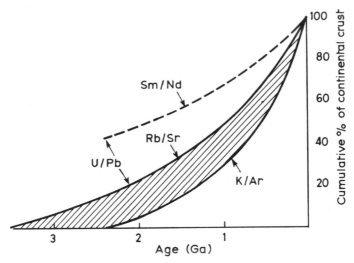

Figure 3.4 Cumulative distribution of continental crust as determined from a variety of radiogenic dating techniques. The biasing toward younger ages increases as the blocking temperature decreases for the different elemental systems. Sm/Nd coupled with U/Pb is the most resilient of the systems, with the highest blocking temperatures, and therefore the upper curve most accurately describes the age distribution of continental crust. The average age seems to be about 2 Ga, but does this reflect recycling of a steady-state volume of crust, like that of oceanic crust though at a much slower rate, or a volume of crust that is not recycling but is gradually growing? Of course, the answer could be some combination of each of these alternatives. (Data from Veizer and Jansen (1984))

to provide a preliminary plot of the ages when rocks were first placed into the crust.

From these data it is possible to determine recycling rates. Some geologists propose that continents maintain a near constant volume from one recycling episode to another, as is the case with oceanic crust; alternatively, continents may experience some net growth; that is, during recycling, the amount of material moving into the continental domain is greater than the volume of material moving out of the continental domain back into the mantle. The Sm/Nd data show that the average age of continental crust is approximately 2 Ga. Does this mean that the crust has maintained a steady-state volume with old crust recycling into the mantle at a rate equal to the generation of new crust? A recycling rate of 2.6 km^3/yr since about 3.8 Ga would produce this curve. Or is there no recycling of crustal rock, and the Sm/Nd curve is an effective rendering of the growth history of the continental crust? The answer probably involves some combination of both, a limited amount of recycling and a growth history that follows one of the growth curves of Figure 1.10.

The two principal arguments favoring a steady-state volume, with recycling of continental crust, involve: (1) characteristics of the continental freeboard, and (2) something called ε_{Nd}, the normalized ratio of the radiogenic ^{143}Nd to primordial ^{144}Nd.★ These two topics are discussed before presenting the more favored model involving a continental-growth scenario. If this is too confusing you may wish to move forward to the conclusions.

Continental freeboard

Like the 'free' board of a yacht that lies above the water line, continental freeboard is the mean altitude of land above sea level. The freeboard of continents is generally believed to have remained constant throughout most if not all of the geologic record. This is in part because no systematic relationship exists between stratigraphic age and altitude of ancient shoreline sequences, although this argument is difficult to quantify owing to problems regarding differential rates of erosion and tectonic processes. Because sedimentary deposits are subject to erosion and intracrustal recycling, examples of nearshore lithologies are increasingly fewer in progressively older stratigraphic intervals. Nevertheless, with the existing stratigraphic record that is biased toward younger ages, ancient shoreline deposits present no relationship between stratigraphic age and altitude of outcrop. A more rigorous data base supporting a uniform freeboard involves the extent of land that has been flooded during the major marine transgressions of the Phanerozoic; because these areas are roughly equal, it must follow that the mean elevation of continents has not changed during the past 600 m.y. (Figure 3.5).

Hence, it is feasible that sedimentary deposits have had the same base level since the Archean, plus or minus the several hundred meters of fluctuation due to glaciation and short-term changes in midocean-spreading rates. The proponents for no continental growth therefore argue that the constancy of freeboard requires that the ocean basins as well as the volume of seawater have remained fairly constant for at least the past 2 to 3 b.y. This argument stems from the conclusion that if continents were growing, a major thesis of this book, the volume of ocean basins would have had to shrink, and assuming a near-constant volume of seawater,

---

★$\varepsilon_{Nd} = [(^{143}Nd/^{144}Nd(\text{sample})) - (^{143}Nd/^{144}Nd(\text{meteorite}))] \div (^{143}Nd/^{144}Nd(\text{meteorite})) \times 10^4$. This is one of several ways of expressing ε_{Nd}.

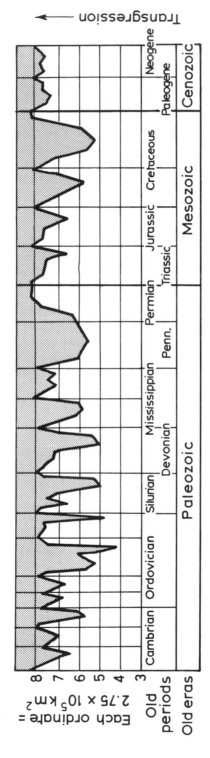

Figure 3.5 Graph depicting the maximum area of the North American continent that was flooded during successive highstands of sea level during the Phanerozoic. Because in each instance the maxima are equivalent, the mean elevation of the continent was constant throughout this 600 m.y. period, and the mean freeboard has not changed for at least this period of time. (Modified from Schuchert (1916), Kuenen (1939), and Burton, Kendell, and Lerche (1987))

shoreline deposits should reflect a continuously rising transgression over continents.

Plate tectonics provides a solution to this conundrum. The putative solution, however, requires that in addition to the constancy of freeboard, three other conditions are assumed: (1) The thickness of continental crust has remained constant since its formation, that is Archean continents were no thicker than continents of today. The bulk composition of continental rocks has not changed systematically with time, and therefore the density has not changed. Hence it follows that the thickness of the crust too has been uniform since the beginning of continental growth. (2) The volume of water in the upper mantle, crust, and atmosphere has remained uniform for the past 4 b.y. The concepts regarding the origin of the hydrosphere changed following the discoveries of the *Apollo* missions. The 4.5 b.y. old samples of lunar rock, which are believed to be products of the Earth's primordial mantle and crust, are essentially devoid of volatiles. Therefore, the Earth probably acquired most of its volatiles after the main stage of solar-system accretion. Thus, seawater is a product introduced to the Earth by impacting planetesimals; the volatile water molecules were collected during the 500 m.y. of intense bombardment after the moon had split off from the Earth (Figure 1.2). (3) The surface area of the Earth has remained constant, that is the Earth has not been expanding. The physics of angular momentum demands this conclusion.

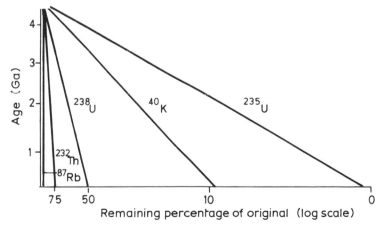

Figure 3.6 Semilog plot showing the loss of radioactive material since the formation of the Earth. To determine the effective cooling of the Earth, these data need to be integrated with the actual abundance of each element in the crust–mantle domain.

If one accepts the general accuracy of these inferences, then midocean-ridge geometries can be reconstructed which will counterbalance the effects of an increasing area for the ocean-basin floor. Because the Earth has been cooling with time as a result of diminished abundance of radioactive substances (Figure 3.6) and because much of the heat that is generated from the Earth is purged along the ridge systems, it follows that the combined effects of ridge length and spreading speed would necessarily diminish with time. The apparent record of a uniform freeboard provides an intriguing method for the reconstruction of the requisite volume of the global ridge system as a function of continental volume. Figure 3.7 portrays ridge profiles corresponding to a sequence of spreading speeds. The crustal subsidence away from the ridge axes is a consequence of thermal decay which results in an increasing density within the oceanic lithosphere. These profiles are based on the conditions dictated by the modern geothermal gradient. The increased incidence of komatiites (ultramafic lava) in

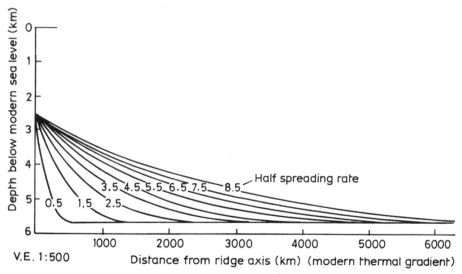

Figure 3.7 Theoretical profiles as a consequence of the thermal decay of oceanic crust of one side of a midocean-ridge system, spreading at various rates. In general, active ridge crests are about 2500 m below the sea surface, and the lithosphere subsides to its near maximum of about 5500 m during the first 80 m.y. Rates shown are 'half-spreading speeds', the velocity relative to a fixed point of the crest and not to a point of the complementary ocean crust that is spreading in the opposite direction. These data clearly show how the volume of the ocean basins is progressively diminished with increasing rates of spreading; see Figure 3.8. (Data from Sclater, Parsons, and Jaupart (1981))

Archean stratigraphic intervals suggests that, at least locally, heat-flow gradients were higher than modern gradients. Komatiites crystallizes at about 1600° C instead of 1200° C, the approximate crystallization temperature of normal MORB. If this were the case, then the amount of ridge inflation could have been greater. Thus the values of ridge widths are minimum values (Figures 3.7 and 3.8), while the lengths and spreading

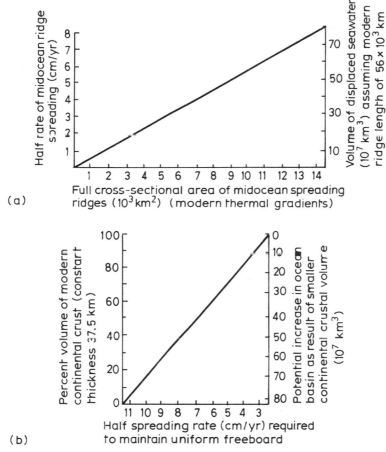

(a)

(b)

Figure 3.8 (a) Graph depicting the reduced volume of the ocean basins as a function of different spreading rates. The length of the ridge system is fixed at the modern length of 56 000 km, and it assumes a modern thermal gradient with a crustal subsidence following the decay curves shown in Figure 3.7. (b) Graph depicting the spreading velocities of a 56 000 km long midocean ridge system required to maintain a constant continental freeboard for various volumes of continental crust and the correspondingly larger ocean basin. The graph supposes an Earth of fixed size with a constant volume of seawater of $137 \times 10^7 \, km^3$.

rates shown in Figure 3.10 are maximum values. If all this seems like a house of cards, let me point out three additional observations that bolster the logical framework: (1) komatiites have been found on the modern Galapagos ridge system and this ridge system is not anomalously broad or inflated; (2) metamorphic mineral zonations have the same spatial relations in Archean settings as in Phanerozoic settings, suggesting similar pressure–temperature gradients then as now; and (3) because of the large lithostatic pressure necessary for the formation of diamonds the occurrence of Archean diamonds requires that the thickness of continental lithosphere and the depth of the 1400° C isotherm has remained at about 150 km. The argument continues as follows. Because of the constancy in the thickness of continental crust (37.8 km), a linear relation exists between continental crustal volume and area of ocean floor (Figure 3.8(b)). Therefore, if continents have been growing since 4 Ga, the area of ocean floor must have been shrinking. But if the volume of seawater has not changed since 4 Ga, then the average height of the seawater column must have systematically decreased in order to maintain a constant freeboard (Figure 3.9, Table 3.1). How can this happen?

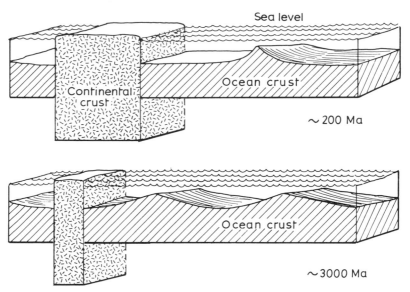

Figure 3.9 Schematic diagram illustrating how with a constant volume of seawater a constant continental freeboard can be maintained by systematically diminishing the length and volume of the midocean ridges in order to adjust for the increasing volume and area of continents.

Table 3.1 Dimensional statistics for the continental crust and oceanic basins following incremental reductions (intervals of 10%) of the continental crustal volume. The far right-hand column provides the volume of ridge crest *in addition* to the volume of modern ridges necessary to maintain a constant freeboard.

Continents			Oceans			
Volume *(10⁹ km³)*	*Area* *(10⁶ km²)*	*Thick-* *ness* *(km)*	*Water* *volume* *(10⁹ km³)*	*Area* *(10⁶ km²)*	*Height* *(km)*	*New* *ridge* *volume* *(10⁶ km³)*
7.60	201	37.8	1.37	309	4.43	–
6.84	180	37.8	1.37	330	4.15	91.9
6.08	161	37.8	1.37	349	3.93	176.1
5.32	140	37.8	1.37	370	3.70	269.1
4.72	125	37.8	1.37	385	3.65	335.6
3.96	105	37.8	1.37	405	3.38	424.2
3.04	80.4	37.8	1.37	430	3.19	534.9
2.43	64.2	37.8	1.37	446	3.07	605.8
1.52	40.2	37.8	1.37	470	2.91	712.1
0.76	20.1	37.8	1.37	490	2.80	800.7

By lengthening midocean ridges and increasing spreading speeds of ridges, the average depth of the ocean basins will decrease. For example, during the interval of 2 to 3 b.y. ago, when the volume of continental crust was possibly 50% of its present volume, constancy of freeboard may have been achieved by any of the following processes (read from Figure 3.10): ridges spreading at today's rate with a total length of 162 000 km (the modern 56 000 km plus the additional 106 000 km); ridges spreading 1.8 times as fast (4.5 cm/yr, half rate) with a total ridge length of 114 000 km; ridges spreading 2.6 times as fast as today with a total length of 96 000 km; or ridges spreading at 8.5 cm/yr (half rate), 3.4 times as fast as the recent average, with a total ridge length of 86 000 km.

With larger ocean basins, longer ridge systems are a reasonable expectation; so also, with the larger area of oceanic crust, higher spreading rates could be accommodated because the additional space between ridges and subduction zones allows for cooling and the required increase in density of the crust that are necessary to provide the negative buoyancy – slab-pull – forcing function. The shaded region of Figure 3.10 is an envelope that may bracket the actual kinematic parameters of plate tectonics. This reflects a plate tectonic history involving a gradual shortening in length and slowing in spreading of the midocean-ridge

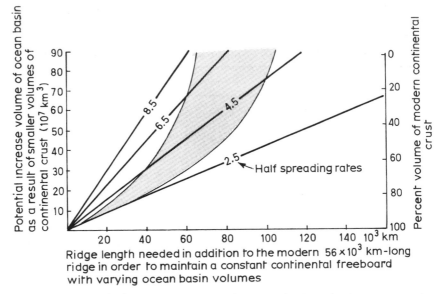

Figure 3.10 Graph depicting various midocean-ridge lengths that are spreading at various speeds needed to compensate for smaller volumes of continental crust and a correspondingly larger volume for the ocean basins. A constant freeboard and uniform volume of seawater is assumed. The shaded region is an intuitive guess of what conditions may have prevailed during the Earth's history where the continents have been growing and the Earth has been cooling.

system. At the same time there would be a concomitant reduction of the length and rate of consumption along subduction zones; therefore, the volume of arc volcanism, and consequently the rate of continental growth, may also have diminished through time; see line DGH of Figure 1.10 for a rendering of such a possible growth pattern.

ε_{Nd}

Within the mantle and crust ^{147}Sm decays to ^{143}Nd with a half life of 1.063×10^{11} yr; thus, the value of ε_{Nd} steadily increases with time. As partial melts fractionate from a mantle reservoir, however, rare-earth elements are relatively enriched inversely proportional to their atomic number. The mantle therefore becomes relatively depleted in Nd, while the crust becomes relatively enriched (Figure 3.11). Analytical measurements of Nd isotopes indicate that crustal values are not as enriched in ^{144}Nd as is predicted from light rare-earth patterns (LREE), and this has indicated to some researchers that crustal material must be recycled back into the mantle; in effect, the depleted reservoir is enriched back toward a more

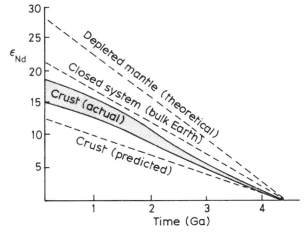

Figure 3.11 Graph depicting the systematics of ε_{Nd} for the bulk Earth in a closed system. Top line represents a depleted mantle where the lighter Nd is preferentially lost to the crust in relation to the heavier Sm. The lower line indicates the predicted complementary relation for the enriched crust, but note that the actual measured values are less enriched than expected. Hence, the mantle is somehow being enriched in Nd; perhaps crustal material is recycled back into the mantle, or perhaps the mantle is stratified, and a lower reservoir of less depleted material cycles into the more depleted upper mantle.

primordial state (Figure 3.11). If 35% of the crust is reintroduced into the mantle every 1 b.y., a mass balance of crustal recycling is produced that fits the observed data. This is equivalent to 2.66 km^3/yr, a flux that exceeds the total amount of sediment cycled on and off the oceanic basement. An alternative solution for the ε_{Nd} data involves recycling between an upper and lower mantle to produce the necessary enrichment of Nd. The known rate of depletion of all isotopes within the mantle suggests that the volume of the reservoir that feeds the crust is restricted to the upper 600 km. The crust is enriched in numerous isotopes as a consequence of chemical fractionation, but the extent of crustal enrichment is so large that the entire mantle could not have functioned as a reservoir; the crustal sink is simply too small to effect depletion of the entire mantle. Consequently, in the nonrecycling model, hotspot plumes rising from the core–mantle interface are envisaged as a way to stimulate limited circulation between the primordial lower mantle and the depleted upper mantle. This intramantle cycling of material could provide the necessary enrichment of ^{144}Nd in the upper mantle to explain the ε_{Nd} values.

Thus, the ε_{Nd} data do not provide a unique solution; the source of the upper mantle enrichment of ^{144}Nd could be either the lower mantle or the crust. The mechanism for the lower-mantle option is rather speculative and therefore may not be favored by some geologists. But the crustal-recycling option is seemingly less attractive because of the large volume of low-density crustal material necessary to be recycled.

Subducting continental crust

In the simplest terms, subduction is the sinking of material into the mantle. Because continental crust has the bulk composition of andesite, with a density of approximately $2.7 \, \text{g cm}^{-3}$, it would seem to be impossible to subduct this crust into the mantle where the density is at least $3.2 \, \text{g cm}^{-3}$. Some geologists have suggested that continental crust that is attached to oceanic crust is pulled into the mantle, but the low tensile strength of the crust makes this a rather dubious possibility. More intriguing is the concept that continental crust, if subjected to conditions of high enough temperature and pressure, metamorphoses to a more dense medium. Laboratory experiments confirm that normal crustal material, when metamorphosed to granulite phases, becomes more dense (from 2.8 to $\approx 3.1 \, \text{g cm}^{-3}$); but this new condition is still not dense enough to allow the crust to sink into the mantle. If the crust were composed entirely of gabbro (pyroxene, plagioclase, $\pm$ olivine $\pm$ quartz $\pm$ spinel) subduction could occur, as this material metamorphoses to eclogite (pyroxene, garnet $\pm$ quartz) with a density of as great as $3.5 \, \text{g cm}^{-3}$. But most of the continental crust is not composed of gabbro. Local occurrences are possible, especially in collision zones where the lower crust would be subjected to the appropriate pressure and temperature conditions, but the subduction of an entire crustal slab on a routine basis seems unlikely. Another unproved mechanism supposed by some geochemists to recycle crustal material into the mantle involves sediment subduction.

Sediment subduction

The quandary of sediment subduction is similar to the sinking of a ping pong ball in a glass of water. Sediments are low in density (commonly $1.6 \, \text{g cm}^{-3}$ for pelagic mud to $2.7 \, \text{g cm}^{-3}$ for dense sandstone); therefore, sediment cannot sink gravitationally into the mantle. Nonetheless, the geochemistry of arc volcanic rocks suggests that small amounts of contamination from oceanic sediments occur. Seismic-reflection data also show that some sediment may ride piggyback on a descending oceanic plate in a subduction, particularly if small tensional basins form when the oceanic crust flexes downward, seaward of the trench (Figure 3.12). But what is the volume of the subducted sediment?

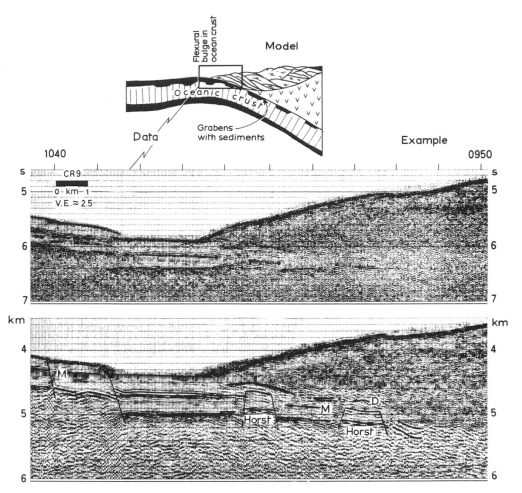

Figure 3.12 Model depicting a flexural bulge on the ocean crust forward of a trench and the resulting horst and graben structures creating 'buckets' in which sediment may pass beneath the deformation front of the accretionary prism and possibly be carried into the mantle. The actual example comes from the Middle America trench off Costa Rica; the top seismic line is an uninterpreted time line, and the bottom is an interpreted depth section. (Data from Shipley and Moore (1986))

Geochemistry

Numerous chemical analyses have been performed in an attempt to characterize the chemistry of the magma source. The ODP project has planned several drilling experiments to sample sediment and ocean basement at the edge of subduction zones in order to determine if differences there explain chemical variations that occur longitudinally along the adjoining volcanic arc. These are 'what goes in and what comes out' experiments. Beryllium-10 and isotopes of lead (particularly $^{207}Pb/^{204}Pb$) are found in some volcanic rocks in quantities exceeding normal MORB and potential mantle sources; therefore, an admixture of sediment is required. But a consensus view, from the community of volcanic petrologists, is that 5% or less of the sediment pile entering a subduction zone is needed to supply the mass of the contaminating elements observed in volcanic arc rocks.

The isotopes ^{177}Hf and ^{176}Lu fractionate from the mantle in a consistent way owing to their similar geochemical behavior in mantle minerals. However, under shallow-crustal and weathering conditions, these two elements fractionate inconsistently; the ^{177}Hf behaves chemically more like zirconium, and therefore this element is enriched relative to ^{176}Lu in zircon.

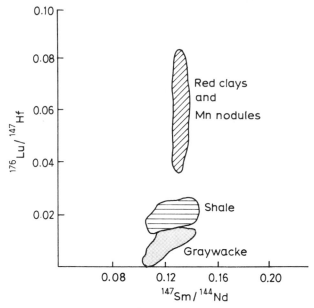

Figure 3.13 A plot of the ratios of $^{176}Lu/^{177}Hf$ against $^{147}Sm/^{144}Nd$ demonstrating the pronounced fractionation of the former and the near-perfect uniformity of the latter. If significant amounts of sediment were subducted – recycled – into the mantle, the chemistry of seamounts or volcanic-island arcs should reflect variations of the above elements because of the uneven distribution of graywacke, shale, and pelagic red clay around the seafloor of the ocean. But such a variation is not found. Either rapid mixing takes place within the mantle domain or little sediment is being subducted. (Data from Pachett *et al.* (1984))

Zircons, because of their high specific gravity, tend to accumulate in continental margin settings, nearshore to deep-sea-fan deposits. Because Lu is not concentrated in any detrital or upper-crustal mineral, Lu becomes enriched relative to Hf in deep-sea clays, the depositional sink of last resort. Thus, ^{177}Hf is partitioned physicochemically relative to ^{176}Lu within the surfacial domain of the crust (Figure 3.13). If the sediment lying on the oceanic crust was subducted at rates approaching 1.5 km^3/yr, inhomogeneities should exist within the mantle between these two elements. Thick accumulations of terrigenous debris, laden with zircons, form isolated patches about the globe: Amazon cone, Indus and Bengal fans, and the thick accumulations throughout the Indonesian archipelago are the principal areas; whereas thin deep-sea-clay accumulations, enriched in ^{176}Lu, are more evenly and widely disseminated. Thus, the relative abundance of ^{176}Lu and ^{177}Hf within subduction zones is expected to be variable, controlled by the nature of the sediment lying on the ocean crust. Despite the spatial separation of the two elements, no igneous bodies extruded from the mantle reflect a relative enrichment of either element.

Some authors have argued that the subduction of sediment is a selective process, with the coarser-grained trench deposits being accreted and the fine-grained clays lying on the basaltic carapace being subducted. The systematics of ^{176}Lu and ^{177}Hf implies that selective subduction is not a significant phenomenon, and the local enrichment of the short-lived ^{10}Be requires that the youngest sediments are somehow involved in magma contamination. These data seem to be contradictory, and they neither help understand nor constrain the amounts of sediment being subducted.

Seismic reflection profiling (Stratigraphy)

The art of seismic-reflection profiling has progressed rapidly during the past decade. Seismic profiles across subduction zones provide clear pictures of the fate of the incoming sediment. Figure 3.14 illustrates an example from the Nankai Trough. Here, the Philippine plate is being subducted beneath Japan. The sediment in the Shikoku basin consists of a layer of clay and associated hemipelagic material at the base. This blanket of fine-grained mud is overlain by coarser-grained submarine fan material along the continental margin. The oceanic trench represented by the Nankai Trough is essentially filled with layers of turbidites and slump deposits. The initial phase of sediment accretion begins along the leading edge of the overriding plate.

ODP drilling into the accretionary prism east of Barbados at the Lesser Antilles arc penetrated a structural discontinuity between the deep-sea mud and the submarine-fan deposits. This decollement is a result of dewatering of sediment as it compacts within the collision zone, between the arc massif

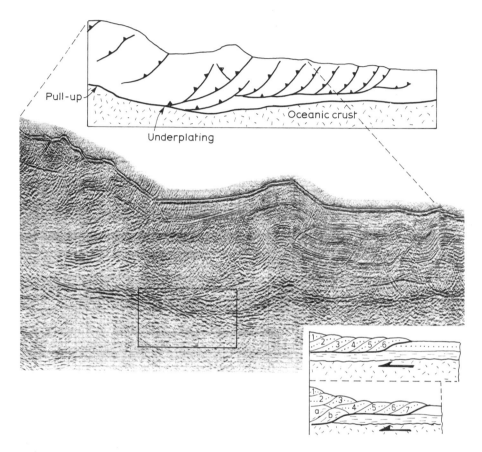

Figure 3.14 Seismic-reflection time line across the deformation front of the Nankai Trough, south of Honshu Island, Japan. The initial decollement is within the pelagic layer of deep-sea mud. Following subduction of the pelagic material, the decollement ramps downward to a surface of pronounced velocity impedance, probably on or within the basalt layer of ocean crust. In this manner the pelagic mud underplates along the lower surface of the accretionary wedge. Note that the apparent rise of the basalt surface toward the left, arcward, is an artifact in the seismic-reflection time section. This 'pull-up' results because of the faster velocity of sound in basalt than in the surrounding sediments. The surface flattens or even dips slightly downward after converting the image into a depth section, where the spacing between reflection horizons is adjusted for the velocities of sound in the varying stratigraphic media. (Data from Taira, Tokuyama, and Soh (1989))

and the incoming oceanic plate. Along the decollement, the hydrostatic pressure builds to a level equal to the lithostatic pressure. Under these circumstances, the strata above the decollement are essentially floating, and the underlying material slips unimpeded toward the arc. If the decollement persists at this stratigraphic interval, the deep-sea muds will be subducted into the mantle. But fluid pressures are unstable, and the decollement may slice down through the mud layer onto or within a discontinuity associated with the basaltic layer of the oceanic crust. In this fashion the mud layer, and possibly a thin layer of basalt, becomes underplated onto the base of the accretionary prism; only the igneous oceanic crust sinks deeper into the mantle.

In several places subduction zones have been imaged that seem to be impoverished in accreted sediments, for example along the Japan Trench east of Hokkaido, nearly all of the Mariana arc, and at parts of the Guatemalan and Peruvian margins along the eastern Pacific. Furthermore, DSDP and ODP drilling results indicate distinct episodes when these continental margins rapidly subsided, locally by as much as two kilometers. Some investigators argue that these two observations, little accreted sediment and episodes of crustal subsidence, require a phenomenon called subduction erosion. Thus, sediment as well as a portion of the continental crust may have been subducted. These are convincing data, and the circumstance remains perplexing. How does low-density crustal material sink into the denser mantle? Could the structural truncation involve either a lateral transfer by strike-slip faulting or an arcward transfer followed by underplating (subcretion), analogous to the mud layer of Figure 3.14?

Modeling the mass balance of subduction zones

Another potential means to determine if or how much sediment is subducted is to calculate an accretion efficiency factor for subduction zones; that is, to model the amount of sediment that has entered a subduction zone and compare this with the amount of sediment that has been accreted based on detailed seismic surveys. The major problem with this approach is reconstructing the amount of sediment that has entered a subduction zone. Figure 3.15 displays three generally east-trending profiles across the Lesser Antilles accretionary complex. This system has been active for approximately 50 m.y. and involves the subduction of a thin and rather uniform layer of deep-sea clay and a thicker sandy unit that is quartz rich, implying a continental source. The sandy unit that currently lies seaward of the trench is thickest in the south (about 4 km) and thins toward the north (< 1 km). The geometry of the unit and the petrography of the sands suggest a southern source. It is believed that much of the accretionary prism and the remaining amount still on the Atlantic seafloor are part of a submarine fan complex sourced by the Orinoco River of Venezuela. The fracture zones, Tiburon

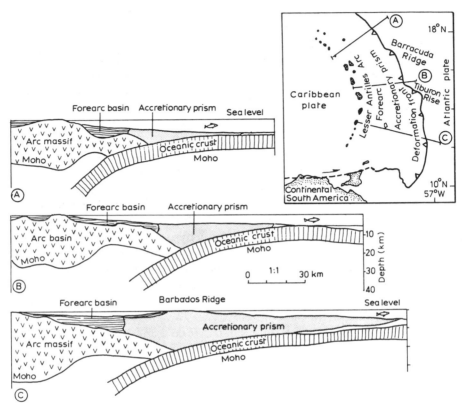

Figure 3.15 Line drawings from three seismic lines that run transverse to the Lesser Antilles island-arc system. Note the systematic increase in the volume of the accretionary prism from north to south. The volume of material on the incoming Atlantic seafloor also thickens toward the south. The principal source terrain for the detrital material is the South American continent. Besides the normal thinning away from a source terrain, the east-trending Tiburon Rise and Barracuda Ridge have restricted the northward flow of sediment. These edifices are related to transform faults on the North and South America plates and therefore probably extend far to the west on the part of the plate that is now subducted. Thus, the present expression of thickness on the incoming plate is a reasonable indication of the nature of sediment being subducted since the arc began at about 50 Ma. The correspondence between thickness of incoming sediment and cross-sectional profiles of the accretionary prism lends credence to the supposition that little sediment is being subducted. (Data from Speed and Westbrook (1984) and others)

Rise and Barracuda Ridge, have restricted the northward transport of the sandy turbidites. Seaward of the trench the incoming pile is about 1.5 km thick just north of Tiburon Rise and about 0.8 km thick north of Barracuda Ridge. It may be impossible ever to know the original volume of the Orinoco fan, but the shape of the Lesser Antilles accretionary prism shows an excellent correlation between the three thicknesses of the incoming pile from north to south and the matching cross-sectional area of the accreted prism (effectively the same as a volume factor):

	Thickness (km)	Area (km²)
North	0.8	410
↓	1.5	855
South	4.0	2520

A similar, but more detailed, analysis is possible for a segment of the Aleutian accretionary prism. Figure 3.16 is a line drawing from a processed seismic profile that trends north across the arc in the midregion of the Aleutian chain near Amlia Island. The polygon of low-velocity material is interpreted to be accreted sediment resulting from a combination of offscraping and underplating processes. Its cross-sectional area is 590 km².

The Aleutian arc began to form approximately 55 m.y. ago. Figure 3.17 characterizes the nature of the three principal phases of subduction. For the period 55–43 m.y. ago, the Late Jurassic oceanic plate was subducted at a rate of 20 cm/yr. The Pacific–Kula spreading ridge lay approximately 3500 km to the south, and the age differential between the incoming plate and the arc averaged 100 m.y. Because the plate was remote from any continents, only pelagic sediment would have been on the oceanic crust. With a pelagic rain typical for high latitudes, 1 mm/1000 yr, the cross-sectional area of sediment available for accretion during the first 12 m.y. would have been

$$0.1 \text{ km} \times 200 \text{ km/m.y.} \times 12 \text{ m.y.} = 240 \text{ km}^2.$$

The kinematics of plates were abruptly altered 43 m.y. ago, presumably as a result of collision tectonics between India and the Eurasia plate. These plate reorganizations are clearly indicated by the sharp bend in the Hawaiian–Emperor seamount chain. The rate of subduction of the Pacific plate slowed to 7 cm/yr, and the Pacific–Kula spreading ridge became inactive following this plate reorganization. Because the Kula ridge had been moving northward, the age differential was reduced to about 50 m.y. between the incoming plate and the arc, thus bringing about the

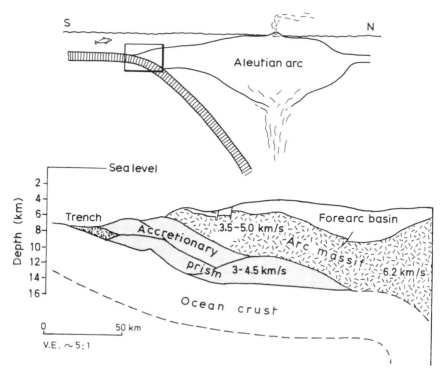

Figure 3.16 Line drawing from seismic-reflection data across the accretionary prism of the Aleutian arc near Amlia Island, central part of the arc system, 173° W. The shaded area of the lower cross section shows low-velocity material inferred to be accreted and underplated sediment. (Data from McCarthy and Scholl (1985) and Scholl and Ryan (1986))

diminished thickness in the pelagic layer. Prior to 5 m.y. ago, only high-latitude pelagic sediment blanketed the oceanic plate. The cross-sectional area of sediment available for accretion during this 37 m.y. period would have been

$$0.05 \, \text{km} \times 70 \, \text{km/m.y.} \times 37 \, \text{m.y.} = 1295.5 \, \text{km}^2.$$

A second distant collision event affected the accretionary history of this remote part of the Aleutian arc. In the area of southeastern Alaska, a sliver of continental crust, the Yakutat terrane, had been moving northward along the Fairweather–Queen Charlotte transform fault system. About 5 m.y. ago this block began ploughing into the North American plate effecting the dramatic uplift of the Alaskan Range. This high topographic relief, augmented by alpine glaciation, brought a flood of sediment into the

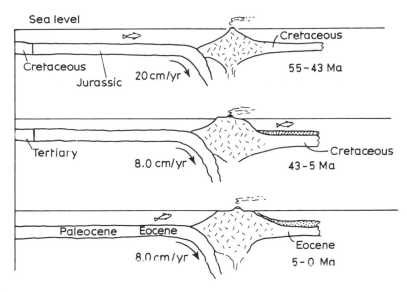

Figure 3.17 Three diagrammatic line drawings across the Aleutian arc system for the area near Amlia Island depicting three phases of subduction during the 55 m.y. history of this arc system. The first phase involved rapid subduction of a relatively old oceanic plate. Subduction slowed following the reorganization of plate motions at 43 Ma, indicated by the abrupt bend in the Hawaiian–Emperor seamount chain. The average age of the incoming plate was also younger than the oceanic crust during the initial phase of subduction. Until 5 Ma the trench in this remote part of the arc was probably empty, but after this time the trench filled to at or near its present state as a consequence of glaciation and the rapid uplift of the Alaska Range several thousand kilometers to the northeast. The trench fill is composed of quartzofeldspathic turbiditic material. An estimate of the total volume of incoming sediment is only slightly greater than the volume of sediment believed to be incorporated into the accretionary prism. (Data from Scholl, Vallier, and Stevenson (1987))

ocean. The long and essentially empty Aleutian Trench became an effective conduit for turbidity currents cascading off of the Alaskan shelf. Thus, the Aleutian Trench was filled with coarser-grained sediment. In the region of the line drawing (Figure 3.17), the trench contains 1.8 km of sediment, and a similar amount is assumed for the duration of this new phase. Therefore, for the final episode of accretion, the cross-sectional area of sediment available for accretion would have been

$$1.8\,\text{km} \times 70\,\text{km/m.y.} \times 5\,\text{m.y.} = 630\,\text{km}^2.$$

In all, the cross-sectional area of sediment that moved into this subduction zone amounted to $999.5\,km^2$. In order to compare this area with the area of accreted material we must adjust for the compaction that follows accretion and burial. The 3.0–4.5 km/s acoustic velocity recorded for the zone of accretion suggests a porosity approaching 0%, whereas the average porosity (voids filled with water) of the incoming sediment is likely to average 40%. By reducing the area of the incoming sediment by 40%, a near-perfect accretion efficiency is indicated:

$$599.7\,km^2 \text{ potential vs } 590\,km^2 \text{ actual.}$$

From examples provided by the long-term record at the Lesser Antilles and Aleutian accretionary prisms, sediment, at least locally, seems to be efficiently transferred from the lower subducting plate to an upper plate of the arc massif. These examples combined with the geochemical constraints provided by the partitioning of ^{176}Lu and ^{177}Hf indicate that large quantities of fine-grained sediment are not preferentially subducted. In those instances where subduction zones have a poorly developed accretionary prism or show evidence for episodes of sudden subsidence, some form of tectonic erosion and sediment subduction is implied, although the subducted material has not necessarily passed into the mantle. It may be deep beneath the forearc basement, at present hidden from both the borehole and seismic profile. Nonetheless, opinions vary on this matter. The truncated portion of the forearc, owing to its buoyancy, seems unlikely to be subducted into the mantle. But at arcs such as the Mariana, we must still ask, where has the sediment gone?

3.2 A GLOBAL BUDGET

The global average for the growth of volcanic island arc massifs is estimated to be $1.1\,km^3/yr$. The production of basaltic seamounts is $0.2\,km^3/yr$, and the amount of volcanic ash on the seafloor is equivalent to another $0.05\,km^3/yr$ of newly generated volcanic rock. Sediment on the ocean floor represents continental denudation. Estimating the rate of denudation is possible by converting the estimated $1.35 \times 10^8\,km^3$ of sediment that lies on oceanic crust to a bulk density of $2.75\,g/cm^3$, equivalent to upper-crustal densities, and dividing this by 55 m.y., the average age of the ocean crust. The resulting estimate for continental denudation is $1.65\,km^3/yr$ (Figure 3.18). Thus, the amount of sediment falling onto ocean crust combined with the amount of newly generated volcanic rock indicates an approximate $3.0\,km^3/yr$ flux of rock and sediment; therefore, this amount must enter the global array of subduction zones and become available for accretion. If everything is accreted, net growth of continents would equal $1.35\,km^3/yr$, whereas with an accretion efficiency of 80% the net growth potential for continents would be $0.65\,km^3/yr$ (Figure 3.18).

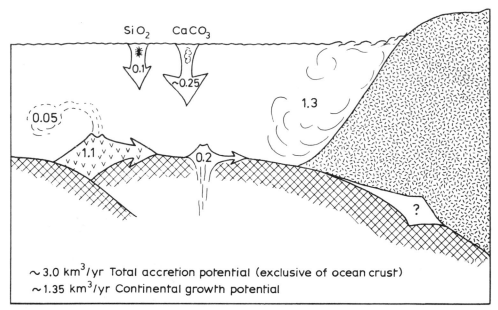

SiO₂   CaCO₃

~3.0 km³/yr Total accretion potential (exclusive of ocean crust)
~1.35 km³/yr Continental growth potential

Figure 3.18 The total volume of sediment on the ocean floor is equivalent to a volume of rock (i.e. the more dense source material) of about $94 \times 10^6 \, km^3$. Dividing this by the average age of the ocean floor provides an average yearly erosion rate for continents. Combining these data with global estimates for the rate of crustal growth for volcanic arcs and seamounts provides an accountant's spreadsheet for crustal-growth potential and continental-crust denudation. The roughly $20 \, km^3$ of oceanic crust that forms each year is not included, as essentially all of it is recycled back into the mantle. Hence, approximately $3 \, km^3$ of material is available for accretion each year, and of this $1.35 \, km^3$ constitutes a net growth potential for continents. The uncertainty of the actual extent of continental growth centers on not knowing the amount of material that is subducted. (Data from Howell and Murray (1986))

The accuracy of these accretion and continental growth estimates are difficult to assess. But one test does support the general magnitude of these estimates. Much of the circum-Pacific, as well as the mountainous region extending from Indochina to the Iberian Peninsula of southwest Europe, is characterized by orogenic belts formed by accretion tectonics. This will be discussed in greater detail in the next chapter. Figure 3.19 depicts the area of accretion since the early phases of the breakup of Pangea 200 m.y. ago. Incorporated in the collage of terranes are blocks of older continental material, for example fragments of the Cimmeria stretching from Turkey to Burma, Sundaland in Indonesia and the Philippines, Omolon in eastern

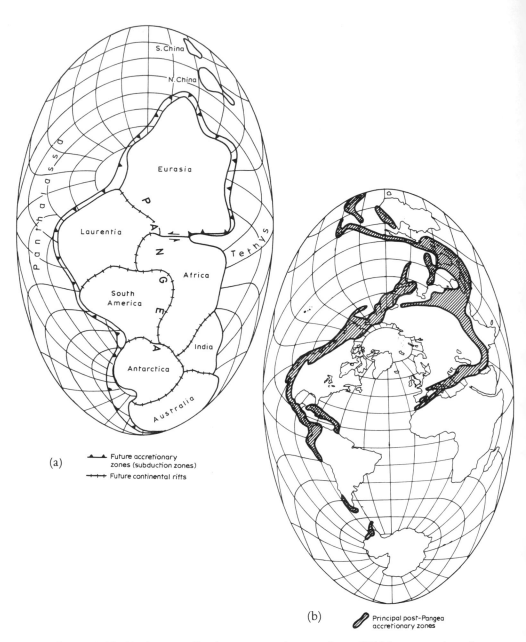

Figure 3.19 (a) A generalized reconstruction at about 200 Ma portraying the supercontinent of Pangea and the surrounding oceans of Panthalassa and Tethys, both of which have been completely subducted. The future sites of subduction and continental rifting are indicated. (b) The present distribution of continents, 200 m.y. after the breakup of Pangea, and the orientation of the two main orogenic belts which record the consequence of the subduction of the oceans of Panthalassa and Tethys.

Russia, Seward Peninsula and the Brooks Range in Alaska, the Alexander terrane in British Columbia, and most of West Antarctica. The mass of these terranes represents relocated crust, and therefore this mass is not included in the aforementioned recycling calculations.

The area exclusive of the obvious pre-Jurassic continental fragments is approximately $25 \times 10^6 \, \text{km}^2$. Inferring an average crustal thickness of 25 km and a period of accretion lasting 200 m.y., we obtain a yearly accretion rate of $3.13 \, \text{km}^3/\text{yr}$. Thus, despite the uncertainties of the generalizations, the estimate of crustal accretion since the breakup of Pangea is consistent with a yearly rate of accretion which is based on the rates of continental denudation and the rate of the magmatic growth of island arcs and seamounts. Because much of this accreted material is recycled continental crust, the effective net growth of continents, at least during the past several hundred million years, is approximately $1 \, \text{km}^3/\text{yr}$. Remembering Figure 1.10, we see that most continental-crust models envisage higher initial growth rates. The slope of line DGH describes a rate of growth of about $2.8 \, \text{km}^3/\text{yr}$ for the Archean, about $1.8 \, \text{km}^3/\text{yr}$ for the next 1 b.y. of the Proterozoic, and a rate of approximately $1 \, \text{km}^3/\text{yr}$ for the past 1.5 b.y. These speculative rates require further testing, but they provide a framework in which to appraise many tectonic effects which are germane to mountain-building processes and to concepts involving continental-growth scenarios.

3.3 CONCLUSION

The geologic literature is replete with numerous geophysical, geochemical, stratigraphic, and petrologic data and models that address the questions of whether or not continental crust, including sediment, is recycled into the mantle. These data and models provide a dizzying array of possibilities and no single observation seems to have a unique solution. The breadth of the arguments has been provided in order to introduce all the complexities. The option of a continuously growing volume of continental crust is the growth model favored in this book, primarily because of the absence of any clear way on how to subduct and recycle into the mantle low-density crustal material. Our opinions of this, however, must remain open.

4

Suspect terranes

4.1 THE RATIONALE

Plate tectonics, largely illuminated by geophysicists with an oceanic data base, provides the continental geologist with a powerful conceptual tool. The application of plate tectonics to orogenic belts led to reinterpretations; the oscillationist ideas involving principally vertical motions of the crust gave way to ideas concerning suture zones and the accretion of island arcs, continental scraps, and oceanic terranes. Asia was recognized to be a fusion of disparate continental blocks, and Alaska a collage of crustal fragments. Flysch basins and volcanic complexes in mountain systems provided basic data from which to infer the kinematics of island-arc accretion. Belts of blueschist and ultramafic rock became the bench marks for suture zones.

Reflecting back on that exciting period, I remember wondering if the early plate tectonicists were going to leave any problems for the next generation of geologists to solve. Paleogeographic reconstructions seemed relatively easy. An industrious student could tackle global problems armed with a few basic ideas such as that andesites indicate island arcs; chert, pillow basalt, and ultramafic rock herald ocean floors; trends in the percentage of potassium record the polarity (geographic facing direction) of volcanic arcs. Geographic complexes had predictable components in predictable sequences, such as the set: volcanic arc–forearc–accretionary prisms. Therefore, limited knowledge of a few key geologic exposures permitted extrapolations to regional settings. Indeed, the first-order problems of mountain building and continental growth were being quickly solved with the new paradigm.

By the mid-1970s, the principal ingredients of the recipe had been identified. But in the 1980s we came to learn that there are also a vast number of minor ingredients and special spices present and the recipe that had been followed was for the wrong bouillabaisse.

If accretion tectonics is the first derivative of plate tectonics, the second derivative is an understanding that the diverse assemblage of lithologic units within a foldbelt—an orogenic system—cannot be assumed to have directly linked genetic relationships. The intrinsic mobility implied by the existence of an orogenic system requires that the geologic history of each unit must be evaluated on the merits of the information gleaned from each outcrop and not from regional generalizations. Continental margins affected by tectonic processes commonly have a region where the stratigraphic elements should be considered *suspect* in regard to paleogeographic linkages both among the elements and between each element and the adjoining continent (Figure 4.1). This does not mean that all the stratigraphic elements are exotic, far-travelled crustal fragments. What is recognized, however, is the possibility – if not the likelihood – that some portion of the region of suspect geology is made up of allochthonous crustal fragments. One should expect such occurrences as a natural consequence of the mobility and transient state of oceanic crust. Thus, the appellation 'suspect' is a warning signal; and when a geologist offers a tectonic analysis that presupposes a fixist stance – where the various

Figure 4.1 Simple rendering of the North American continent showing the outline of the craton, the area of continental crust that has remained relatively stable for the past 600 m.y.; the disrupted craton, a subset of the former but which has been tectonically disturbed during about the last 160 m.y. as a result of compressional thrusting and extensional detachment faulting; and the suspect terranes, the area of crust suspected to have been added to the cratonal area as a result of accretion tectonics during the Phanerozoic (approximately the past 600 m.y.)

stratigraphic units are all facies one to the other – substantive proof must be provided, because this is really no more likely than if the various stratigraphic units are genetically unconnected and their present configuration is due to tectonic mixing. But what are the suspect stratigraphic 'units' or 'elements', how are they recognized, and how are they classified?

4.2 NOMENCLATURE

Any discipline that is fundamentally descriptive in nature has an expanding if not vexing nomenclature. New words must be invented to accompany the evolution of concepts. Hanging new meanings on old words only leads to confusion. Resurrecting and recasting older terminologies often perpetuates aspects of the outmoded ideas. In struggling with the nature of orogenic systems, a variety of concepts – words – have crept into our expanding glossary. A brief review of some of these terms helps to clarify the significance of adopting tectonostratigraphic terranes as the basic descriptive unit for components in an orogenic system.

Microplate

A microplate is a small lithospheric plate. The Indonesian archipelago exemplifies a setting where small crustal plates interact kinematically in a complex fashion (Figure 4.2). Collision-tectonic processes that knead volcanic arcs, sediment-filled basins, seamounts, pieces of continental margins, and young, relatively buoyant oceanic crust characterize such a region. Understandably, similar settings are invoked to explain paleogeographic reconstructions of foldbelts displaying a perplexing array of lithologic units. Microplate tectonics are a viable process creating this kind of an orogenic belt, but the lithologic units that end up in the foldbelt are not complete microplates because the lithospheric underpinnings have been amputated from crustal pieces composing the collage.

Nappe

The term nappe has been in the tectonic literature for over 100 years. The idea was invented originally to explain the conundrum of the 'double fold' of Glarus in the European Alps (Figure 4.3); that is, instead of two recumbent folds that face each other and look like the vertical cross section of the neck of a flaring jar, a single fold was inferred in which the upper limb is detached and displaced horizontally over a wrinkled lower limb. The uses of the terms thrust and nappe vary among modern tectonicists. It is generally recommended to think of a thrust as a contraction fault for which no angle of dip, mode of dynamics, or amount of slip is implied; and

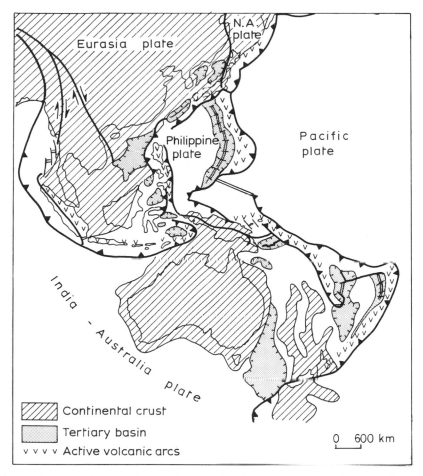

Figure 4.2 Simple rendering of the complex interface between the large plates of Eurasia, India–Australia, and Pacific that extends from New Zealand to Japan. Numerous small plates that to some extent deform in a nonrigid fashion occupy this region.

thrust nappes as allochthonous tectonic sheets that move along thrusts. In some of the older Alpine literature of Europe, a nappe may refer to a thrust sheet of a certain lithology that characterizes a particular paleogeographic affiliation. In this restricted sense, a nappe represents the concept applied to a tectonostratigraphic terrane. But as will become clear below, not all terranes are nappes.

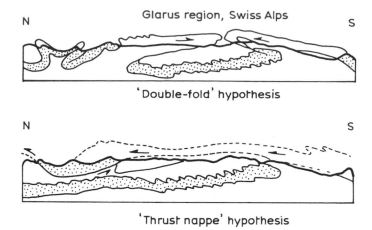

Figure 4.3 So-called double fold of Glarus with Bertrand's late 19th century revolutionary notion of a thrust nappe, which predicts 40 km of crustal closure.

Block

The term crustal block or continental block is a general appellation for crustal material that is allochthonous as in 'the continental blocks of Asia' or in 'blocks in matrix' to describe the fabric of a melange. The term implies a shape and size such that the objects stand out from their surroundings. *Knocker* is a complementary term for outcrop-scale erosion-resistant bodies that emerged from a camp of geologists exploring for chromite during World War II in the melange terrane of California. These workers noted, and I quote from one of the crusty 'oldtimers': 'prominent cone-shaped blocks imbedded in a matrix of mud or serpentinite that protruded out of hillsides and which were mantled with striations indicative of tectonic excitation.'

Flakes

The recognition of oceanic crust tectonically imbedded in strata of a mountain system led to suggested mechanisms of emplacement. The obduction, or transfer from a low-lying seafloor setting up onto the continental margin, was thought to reflect a collision process. The obducted sheets reflect what is sometimes called flake tectonics, which describes the occurrences when two thickened crustal masses collide (Figure 4.4). One of the colliding masses splits at a midcrustal level and seemingly ingests, 'Pac-Man' style, the other. The overriding crustal jaw is the flake, which can be described equally well as a nappe.

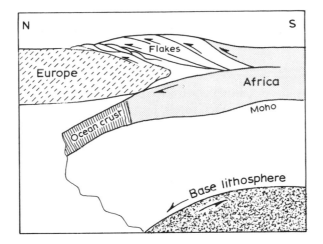

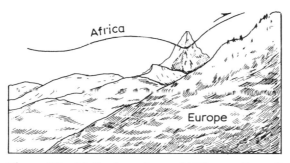

View of the Matterhorn from ski lift near Zermatt

Figure 4.4 A conceptual model depicting the bifurcation of the Africa plate as its upper part collides and overrides part of Europe. The thrust nappes of the Alps are an expression of this collision. The famous peak Matterhorn comprises a slice of Africa at the top, its base is ophiolitic melange representing pieces of the 160 Ma ocean that lay between Africa and Europe, and the granites that surround the town of Zermatt down from the Matterhorn are a part of Europe. This is sometimes referred to as flake tectonics. (Modified from Oxburgh (1972))

Slivers

Just as nappes and flakes indicate planar crustal pieces commonly in a subhorizontal orientation above a thrust fault, slivers are generally vertically standing planar pieces juxtaposed to a transcurrent fault. One person's block could be another's sliver. Flakes, blocks, slivers, sheets, and *fragments* are loosely defined words that prove useful in describing particular settings involving allochthonous pieces of crust.

Province

Cratonal regions of continents are often subdivided into provinces based on contrasting structural characteristics, regional patterns in the age of strata, or geophysical characteristics, particularly magnetic and gravity anomalies. Precambrian crust is often spoken of in terms of provinces, and these provinces can probably be further subdivided into smaller units such as greenstone belts, or even terranes.

Belt

Collision zones and transform margins have a large aspect ratio of length to width; consequently the geometry of zones involving tectonic disruption, accretion, and volcanism almost always results in linear map patterns, or belts. This is a useful term and, as will become clear, a given belt may comprise one or many terranes.

The terms above, in their loosely defined or even colloquial usage, are helpful in describing aspects of an orogenic collage, particularly the structural disruption that is associated with these zones. None of the terms, however encompasses all of the attributes that are necessary to classify the fundamental units that can be mapped in each orogenic system. This is why some workers prefer to think in terms of tectonostratigraphic terranes.

Tectonostratigraphic terranes

Understanding the history of an orogenic belt involves all elements of geology, but primary among these is an integration of stratigraphy and structure. Stratigraphic data provide information regarding the ages and paleogeographic relationships among the assemblages of strata in an orogenic belt. Structural data describe the configuration of these strata. These two data sets provide the means for a tectonic analysis that offers solutions and an explanation for the kinematic and dynamic history.

An orogenic belt is essentially a puzzle made up of a collection of stratal pieces. In order to reconstruct the sequence of events that conspired in creating a given orogenic belt, one must identify the fundamental pieces that compose the puzzle; this necessitates descriptive stratigraphic and structural data. These data define a *tectonostratigraphic terrane*, a fault-bounded package of rock with a distinct stratigraphy that characterizes a particular geologic setting. A tectonostratigraphic terrane is a unique piece in the orogenic puzzle.

Terranes* are generally characterized by their stratigraphy and the implied distinctive paleogeography, but in some cases a metamorphic overprint or tectonic disruption may prove useful in defining the distinguishing characteristics. Thus, terranes can be grouped into three categories: stratigraphic, metamorphic, and disrupted.

Stratigraphic terranes

These terranes are characterized by coherent sequences of strata in which depositionally different relations between successive or adjacent lithologic units can be demonstrated. If crystalline basement rock is within the terrane boundary, its characteristics are helpful in classifying stratigraphic terranes into four broad categories. The geologic history of large terranes may encompass more than one of these categories:

(a) Fragments of continents. These terranes are characterized by the presence of old strata (generally at least as old as the Carboniferous, i.e. pre Pangea) and sedimentologic units reflecting multiple cycles of deposition (quartz-rich rather than lithic-rich sediments). Because oceanic crust gets recycled on average every 100 m.y. and because all oceanic crust older than the Jurassic has been subducted, it follows that any crust older than about 200 Ma must be either continental crust or fragments of oceanic crust that have been emplaced into the continental crustal domain. Crystalline basement is not always preserved in a given terrane, but one should ultimately account for the crustal component of the basement somewhere in the terrane collage. The loss to one continent because of tectonic calving is usually the ultimate gain to another continent, owing to accretion and enlarging of an orogenic collage. The Precambrian crustal blocks composing Asia, the Paleozoic basement terranes of Sundaland, the Hercynian massifs of Europe, and the Precambrian-floored flakes and slivers west of the San Andreas fault in California are examples of continental terranes.

---

*The term *terrane* describes a particular kind of geologic body; therefore a modifier is required for clarification, as in a volcanic terrane or a limestone terrane. A tectono-stratigraphic terrane has both structural (tectonic) and stratigraphic criteria – a fault-bounded package of strata that is genetically unrelated to the adjoining stratigraphic packages – genetically distinct from the other tectonostratigraphic terranes. For brevity one may use terrane without a modifier, but the context must be clear. *Terrain* with this spelling refers to topographic or physiographic features, as in a mountainous terrain, a basin and range terrain, or a desert terrain.

(b) Fragments of continental margins. These terranes are composed of shallow to deep marine clastic lithologies that typically are submarine-fan lithofacies consisting of quartzofeldspathic debris shed from a continental margin. These terranes are commonly massive piles of structurally thickened strata that are associated locally with melange and high-pressure low-temperature metamorphic assemblages. These terranes may mistakenly be assigned to a paleogeographic setting adjoining the continent where they now lie, for example, the Torlesse terrane of New Zealand, the Chugach terrane of southern Alaska, and the Shimanto terrane of Japan. These potentially erroneous interpretations arise because of the great difficulty in distinguishing the petrographic character of continental-margin sediments from one continent to another.

(c) Fragments of volcanic arcs. These terranes are composed predominantly of extrusive volcanic rocks, plutonic roots of arcs, and the sedimentary debris derived from volcanoes. Instances where fragments of volcanic arcs, forearc basins, and subduction complexes occur along a continental margin in an orogenic belt do not prove either that these assemblages formed in their present position or that they are components of the same subduction complex. Some aspects of a volcanic-arc complex are represented in almost all continental-margin orogenic belts; examples include parts of many if not all of the Precambrian greenstone belts, the Paleozoic Avalon terrane of the Caledonides, the island arcs of the Transhimalaya Range in the Tethysides, and the numerous slivers of Mesozoic island-arc terranes throughout the Cordillera of North America.

(d) Fragments of ocean basins. These terranes are characterized by sequences of mafic and ultramafic rocks typical of oceanic crust along with overlying deep-sea sedimentary deposits. Many sequences may have younger continental-margin strata capping the ophiolitic sequences, reflecting either backarc-basin settings or the translational history of the crust from a midocean setting to a continental-margin location prior to accretion. These terranes are generally grossly dismembered. They tend to be confined to tightly compressed suture zones such as the ophiolites of the Indus Yarlung-Tsangpo suture between India and Tibet or the Dun Mountain ophiolite of New Zealand that marks the east edge of the Hokonue terrane. But some ocean-crust terranes occur as relatively undeformed thin sheets (nappes), presumably obducted onto a continental margin, such as the basement of the Dunnage terrane of Newfoundland. In absolute terms, the volume of crystalline ocean basement found in foldbelts is trivial, even though its occurrence is a prominent signal attesting to collision tectonics.

Disrupted terranes

These terranes are characterized by blocks of heterogeneous lithologies and ages set in a matrix of foliated clastic rock or serpentine. Most of these terranes contain fragments of ophiolitic rocks, blocks of shallow-water limestone, knockers of deep-water chert, and bundles of graywacke. In some instances, such as the Central Belt terrane of the Franciscan complex in California or the Anglesey melange terrane of Wales, exotic pieces of blueschist are admixed with these other lithologies. Disrupted terranes have been ascribed to both subduction and transform-margin settings, and the chaotic fabric may indicate a tectonic disturbance, surficial disruption attendant with landslides and debris flows, or stratal mixing resulting from diapiric flow. Debates regarding the causes of the disruption continue to rage about almost every known exposure.

Metamorphic terranes

As a last resort, some rock bodies are grouped as metamorphic terranes. These are characterized by recrystallization and a terrane-wide penetrative metamorphic fabric that has obliterated the original stratigraphy. The Yukon–Tanana terrane (Birch Creek Schist) of Alaska is such a terrane.

4.3 THE MAKING OF TERRANES

The shapes of terranes and the patterns of their distribution reflect one or a combination of several tectonic phases. Terranes begin either by growing intact in an oceanic setting, as in the case of island arcs, or by some calving process that *rifts* a fragment from pre-existing crust. A buoyant piece of crust attached to an oceanic plate will ultimately collide with other buoyant pieces of crust. Where this occurs in an oceanic setting, such as throughout the modern setting of the Indonesian archipelago, the crustal pieces agglomerate or *amalgamate* into larger pieces. In order to distinguish between two similar processes of collision tectonics – the joining of terranes in an oceanic setting and the joining along a continental margin – it is useful to use amalgamation for the former and *accretion* for the latter, even though the tectonic processes of both may be the same. Following all collision events, disruption is likely to continue, resulting in further *dispersion* of the accreted fragments (Figure 4.5). Dispersion could progress into a rifting stage and then a new cycle begins.

Dispersion tectonics in many orogenic belts has dramatically affected the distribution of the accreted terranes. Invariably this will result in uncertainties regarding the number and character of terranes. In the Cordillera of North America, postaccretion strike-slip faulting has the

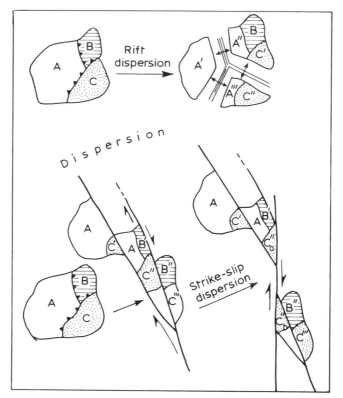

Figure 4.5 Two conceptual presentations illustrating the effects of dispersion. One end member is orthogonal rifting while the other is translational slip. The effect is to produce numerous terranes, and depending on the nature of their respective future geologic histories, the ancient kindred relations may be difficult or impossible to recognize.

effect of dismembering terranes and smearing and scattering the pieces along the margin. Individual fault systems such as the San Andreas, Fairweather–Denali, and Tintina are inferred to have from 300 to 1000 km of slip. The terrane of Wrangellia, where the current distribution of disjunct slivers covers a latitudinal spread of 24°, is a consequence of postaccretion dispersion. The distinctive stratigraphy, particularly the more than 1000 m of Triassic pillow basalt overlain by shallow–water limestone, suggested to geologists that the various pieces may all have been at some time contiguous. This hypothesis has been confirmed by paleomagnetic data, which indicates that all the pieces are displaced to the north, and at the time of magnetization were separated by no more than 4°;

therefore, many of the preliminary subdivisions of the Cordillera that have been identified as crustal fragments of separate terranes were probably once á single entity.

The west-trending orogenic foldbelt that spans from southeast Asia to western Europe, the so-called Tethyan belt, represents the product of a protracted period of terrane accretion and dispersion, principally from the late Paleozoic to the present, along the southeast margin of Laurasia (Figure 4.6). The episode of Triassic and Jurassic accretion involved

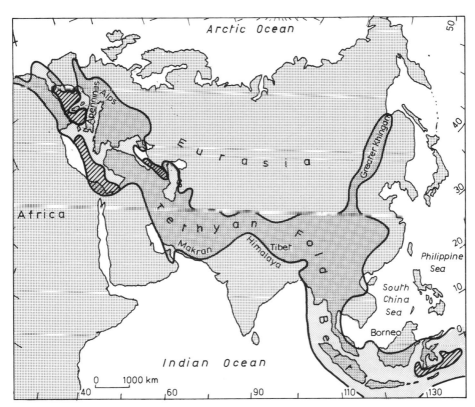

Figure 4.6 General outline of the Tethyan foldbelt that extends from Indonesia in the southeast to the Atlas Mountains of Africa in the northwest. All crust south of the northern limit of this foldbelt is suspect with regard to kindred relations to Eurasia. The belt constitutes a complicated history of both accretion of continental fragments originating either along the northern edge of Gondwana or within the ancient Tethyan realm as continental plateaus, and translational movement of terranes effecting the large-scale dispersion within the belt. The hashed areas represent trapped or juvenal ocean crust. (Modified from Sengor (1987))

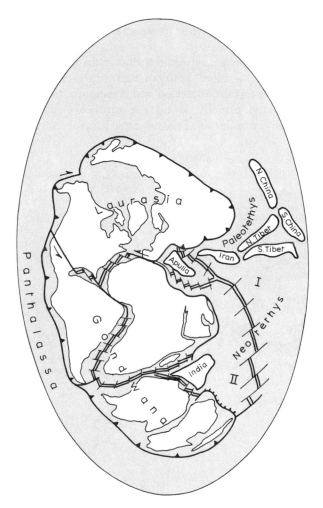

Figure 4.7 Orogenic belts, if they are involved in accretionary tectonics for extended periods of time (50 to 250 m.y.) commonly include the subduction of more than one oceanic plate. Workers in the Tethyan foldbelt often refer to a Paleotethys, the ocean basin that closed in advance of the so-called fragments of the Cimmerian continent, and a Neotethys, the ocean basin subducted in advance of India and the other fragments that accreted since the Late Jurassic. This is rather arbitrary as we do not really know how many oceanic plates may have existed, or the configuration of these plates; there were at least two Neothethys plates (labelled I and II) within the Neotethys Ocean. (Modified from Sengor (1984)). Note that this is a pseudo paleotectonic map; all of the elements shown are believed to be real but the configuration incorporates the salient features covering a time span of at least 100 m.y., from approximately 200 to 100 Ma.

numerous continental fragments, probably pieces from the edge of Gondwana, as well as insular continental blocks that lay somewhere in the expanse of the Paleotethys (Figure 4.7). The accretion collage is sometimes called the Cimmerian continent. But to presume that the current distribution of these agglomerated continental fragments bears any resemblance to the paleogeography of the source regions is unrealistic. The kinematics prior to accretion as well as postaccretion dispersion along the Laurasian margin have certainly disrupted older spatial relationships among the various continental terranes. Superposed on the Cimmerian terranes are a host of younger continental fragments associated with the closure of the Neotethys (Figure 4.7). Thus, the Tethyan foldbelt reflects accretion and postaccretion dispersion as a consequence of protracted subduction along the south margin of Laurasia. Assessing the kinematics of dispersion is hampered because paleontologic and paleomagnetic data, such as used in the Cordillera of North America where transport was largely to the north, are not helpful in ascertaining the sense and amount of longitudinal dispersion toward the east or west. This is a case where it is prudent to identify the terranes in their present coordinates, and not to leap prematurely to conclusions regarding amounts of dispersion and configurations of accretion. The final definitive reconstruction may specify fewer continental blocks as kindred relations than maybe established among a number of the currently delineated terranes.

A second type of dispersion that artificially gives the appearance of an excessive number of accreted fragments involves vertical tectonics. Because the crust is layered, differential movement along faults that traverse all or part of the crust will juxtapose basement rock of contrasting character. In the Coast Ranges of California and the Southern Alps of New Zealand, lower crustal granulite-facies metamorphic rocks are in fault contact with upper crustal granite of Cretaceous age. The granulite-facies rocks were long thought to be Precambrian and possibly of a different terrane. Zircon-age data, however, have shown that the granulites and the granite are different facies of the same magmatic episode. Rapid uplift associated with transpressional tectonics has created this deceptive pattern. Various basement nappes of the European Alps also reflect contrasting lithologies and states of metamorphism. This may be a consequence of stacking of different layers from a single crustal fragment by thrust faulting rather than the juxtaposing of many far-travelled exotic terranes.

A similar scenario is now emerging for some of the Precambrian terranes of eastern Labrador that have been differentiated on the basis of contrasting metamorphic fabrics and mineralogic compositions (Figure 4.8). In this area Archean basement of the Makkovik province (terrane) lies north of the Middle Proterozoic Grenville province. The contact between these two provinces lies within the Trans-Labrador

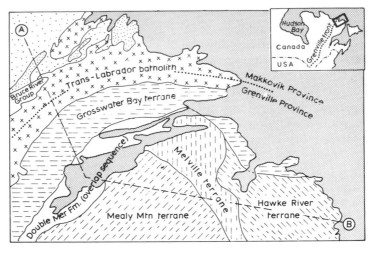

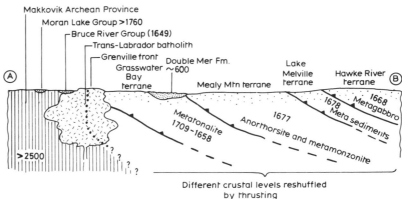

Different crustal levels reshuffled
by thrusting

Figure 4.8 Terrane map along the Grenville front of east Labrador. The nature of
the contact between the Archean Makkovik Province (terrane) and the Middle
Proterozoic Grenville Province (composite terrane) is not known. The contact is
within the Trans-Labrador batholith (TLb); most workers support an intrusive
contact, but a tectonic suture is possible. The TLB must be older than the
1649 Ma volcanics in the overlapping Bruce River Group and younger than the
strata in the Grenville terranes which range from 1658 to 1709 Ma. The oldest
strata to lap across both province are latest Precambrian clastics, about 600 Ma. The
terranes of the Grenville Province constitute distinct fault-bounded rock bodies;
but because of the similarity in ages and the nature of the rocks themselves each
probably represents a different crustal level from a single section of crust. The
dispersion associated with the 1000 Ma Grenville orogeny included reshuffling this
crustal section, but for some reason the TLb was not disturbed. Cross section A–B
is a speculative rendering of the principal tectonic and stratigraphic relations in the
area, but it begs the questions regarding the nature of the Grenville front.
(Modified from Scharer, Krough, and Gower (1986))

batholith. Though the emplacement age of the batholith has not been determined directly, it must lie in the range of 1649 Ma to 1658 Ma, the respective ages for the oldest body to lie on the batholith and the youngest body intruded by the batholith. The strata in the Grenville province are highly disrupted. Thrust tectonics of the approximate 1000 Ma Grenville orogeny has superposed supracrustal metasedimentary strata on deeper crustal metaplutonic rock. These units bear no obvious lithogenetic relationship to one another, but because of similar emplacement ages, determined by U/Pb dating on a variety of accessory minerals from numerous units throughout this province, the Grenville basement terranes and the Trans-Labrador batholith are probably distinct parts of a single crustal block. Crustal segmentation is therefore an aspect of vertical dispersion during the Grenville orogeny. The primary contact between the Mokkovik and Grenville Provinces remains a mystery, which separates an intact Archean block to the north and a dismembered younger crustal block to the south. A tectonic suture involving continent–continent collision would seem obvious, but most published work favors a non-orogenic intrusive relationship.

The foregoing examples have been selected to demonstrate the array of relations that can lead to the formation of terranes, individual fault-bounded pieces of crust that now compose an orogenic collage. By the very nature of this approach, a terranologist should be, in the jargon of classifiers, a 'splitter'. Only after second-order analyses should one make the bold leap to the 'lumping' phase. Many attempts that use plate tectonics to render paleogeographic reconstructions fall into the trap of lumping rock units before the correct identity and genetic history of the pieces is accurately known. Asking how terranes came to be where they are now (accretion, dispersion) involves this second-order phase of analysis. Lumping units together prematurely may mask fundamental differences. One cannot know *a priori* if a specific terrane is wholly exotic, reflecting primarily accretion tectonics, or if the terrane is simply a piece of the local continent that has been dispersed along the margin, either vertically or horizontally. The most rigorous, as well as the most conservative, approach, would be to subdivide a region into as many terranes as possible and then perform experiments to clarify genetic and kinematic linkages. Unfortunately, this only adds to the already burdensome stratigraphic nomenclature, but there is no other satisfactory solution. Studying history requires knowing and understanding all the kings, emperors, caliphs, sultans, etc.

Many classification systems have a hierarchical nomenclature that strives to relate degrees of familiarity. This is not always possible with terrane analysis because of the cyclicity of rifting–amalgamation/accretion–dispersion. When two or more terranes are tectonically joined a

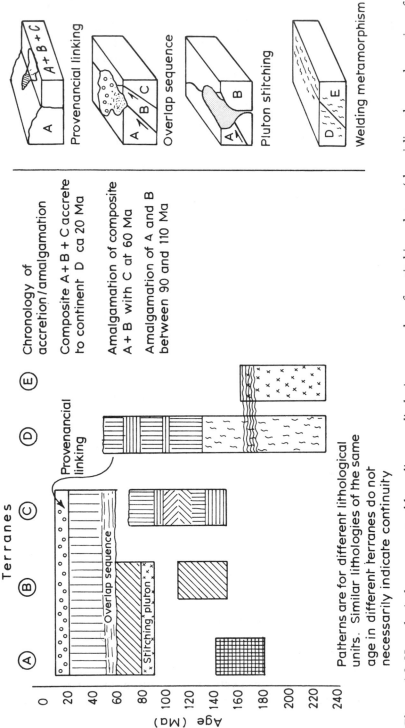

Figure 4.9 Hypothetical terrane assemblage diagram displaying examples of a stitching pluton (chronicling the amalgamation of terranes A and B), an overlap sequence (recording the amalgamation of the composite A + B terrane with terrane C), welding metamorphism (recording latest time that terranes D and E could have amalgamated), and provenancial linkage which indicates when the exotic terranes were juxtaposed to the continental terranes D and E. Note that similar rock types of the same age in different terranes D and E do not constitute binding evidence that the terranes were contiguous.

composite terrane is formed, and the stratigraphy must be stripped back to identify the nature of the pre-amalgamated fragments. But such a composite terrane may rift into two or more pieces or become dispersed along the continental margin. When these fragments join tectonically with other fragments a new list of criteria and new names are required. The problem is similar to a genealogy where a person marries a great grandparent. Where would the offspring of this marriage fit on the family tree? The idea of tracing accretions and dispersions with a classification involving nanoterranes begetting microterranes that beget terranes and onward to superterranes is regrettably not possible. But in the simple case where an area has grown only by amalgamations and accretions, a tectonic assemblage diagram can display this history (Figure 4.9).

Tectonic assemblage diagram

There are five geologic conditions that are useful in reconstructing the assemblage history of a composite terrane or a group of terranes in an orogenic collage: overlap sequences, stitching plutons, welding metamorphism, provenancial linkage and geohistory. These involve a mixture of stratigraphy, petrology, geochemistry, and structural history.

Overlap sequence

When two or more terranes are joined tectonically, there is the possibility that a depositional basin could straddle the suture. In such an occurrence, a stratigraphic horizon will cross depositionally onto both terranes. The basal contacts may or may not be unconformable; this depends on the nature of the tectonics prior to collision. The Paleocene Chuckanut Formation lies depositionally across a collage of terranes in western Washington and southern British Columbia, thus providing a minimum age for the accretion of terranes of this part of North America (Figure 4.10).

Stitching plutons

The accreted terranes may also be associated with plutonic or some other form of igneous activity. Where magma invades a suture a radiometric date provides the minimum age of assembly. If the plutonism can be related unequivocally to the accretion, then the date gives a precise age of amalgamation or accretion. The Chilliwack batholith, a composite body with crystallization ages from 38 to 20 Ma, provides a minimum age for the episode of terrane dispersion along the Straight Creek fault of western Washington and southern British Columbia (Figure 4.10). In Precambrian terranes this technique is particularly useful, not only because of the lack of

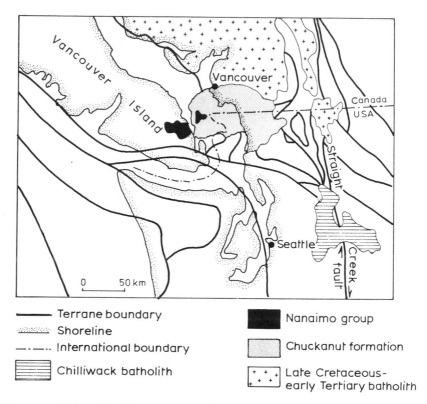

Terrane boundary

Shoreline

International boundary

Chilliwack batholith

Nanaimo group

Chuckanut formation

Late Cretaceous-
early Tertiary batholith

Figure 4.10 Simplified terrane map for southwestern British Columbia and northwestern Washington. The timing for the amalgamation of Mesozoic and Paleozoic terranes is recorded by the approximate 85 Ma Nanaimo Group, a foredeep depositional unit that is incorporated in the terrane collage. The Paleocene Chuckanut Formation is the oldest unit to lie depositionally across the terrane boundaries of the western terranes, therefore fixing the time of accretion to be between 85 Ma and about 60 Ma. The 38–20 Ma old Chilliwack batholith indicates that post-accretion dispersion of the terranes west of the Straight Creek fault system was completed by about 38 Ma.

a fossil chronology, but because dating of 1 to 3 Ga zircons often has a precison to within 1 m.y. Where such dating has been applied to an assemblage of terranes, such as in the Wopmay orogen of northwestern Canada, a detailed sequence of events involving rifting, amalgamation, and accretion has been reconstructed (see Figure 4.19 and the discussion to follow).

Welding metamorphism

Where it is possible to read through regional metamorphism and determine from the protolith that the complex includes more than one terrane, the age of metamorphism provides a minimum age for amalgamation(s). In New Zealand the Jurassic metamorphism that resulted in the Haast Schist records the timing of the amalgamation between the volcaniclastic Caples terrane and parts of the quartzo-feldspathic Torlesse terrane (see Figure 5.14).

Provenancial linkage

Where contacts such as overlap sequences or stitching plutons cannot be mapped directly, the chronology of accretion can be inferred by compositional changes in the clastic population. An element of doubt always remains, however, as unique petrographic fingerprints are rare. The Upper Cretaceous Nanaimo Group, which is part of a collage of terranes cropping out on Vancouver Island and the San Juan Islands to the east, is interpreted as a foredeep deposit. The debris of the deposit collected in a crustal sag in front of stacked thrust sheets is made up of elements of various exotic terranes. The detritus in the Nanaimo Group reflects the composition of the different units within the collage of terranes. Thus, it is possible to reconstruct a tightly constrained chronology for the assembly of terranes in western Washington and southwestern British Columbia as follows: amalgamation during Late Cretaceous time contemporaneous with the deposition of the Nanaimo Group (Santonian Stage, about 85 Ma); accretion prior to the deposition of the Paleocene Chuckanut Formation (60 Ma); and subsequent coastwise dispersion (of the entire package) that terminated before the emplacement of the Oligocene–Miocene Chilliwack batholith (38–20 Ma).

A general test of this technique involves differentiating between times when terranes have a component of oceanic history versus times when terranes are contiguous with the continental margin. This is not a foolproof test, but an abundance of quartz, or a whole-rock geochemistry reflecting high initial ratios (> 0.0706) of $^{87}Sr/^{86}Sr$ for the sandstone and mud fractions of a submarine-fan complex implies contiguity to a continental margin. Many of the California terranes have characteristics which imply a history of coastwise translation rather than transoceanic movement. The chemistry of chert may also provide some insight regarding the spatial relations of a terrane to a continental margin. Chert rich in Al_2O_3 reflects a higher degree of continental proximity than chert rich in MnO and FeO. Combining such information with the radiolarian biostratigraphy permits determination of the rate of approach of an oceanic terrane toward a continental margin.

Geohistory

In concert with the above criteria, paleotectonic reconstructions can be attempted, based on the horizontal and vertical lithofacies that are demonstratively linked in a given terrane. Do these relations indicate continental breakup? Are a volcanic arc and forearc basin preserved? And if so, what is the implied facing direction? In a deep-ocean pelagic sequence does a particularly thick interval indicate an episode of equatorial deposition? Do intervals of coarse-grained deposits suggest earlier episodes of amalgamation? Any clues read from the rocks will help in deciphering the overall tectonic history of terranes in an orogenic collage.

Terrane boundaries

By definition, all terranes must be separated from adjoining terranes by a fault. These fault zones are commonly characterized by a belt of melange, crushed rock, blueschist, or ophiolite, but in many instances terrane boundaries are cryptic, inconspicuous, or simply unimpressive fault zones. Where necessary, a fault or suture may be inferred between areas with different stratal units if the boundary is not exposed and the units on either side cannot be linked by reasonable lithofacies characteristics. This is the conservative tack, mentioned above.

Confusion may arise in discriminating between fault or suture-bounded terranes and fault-bounded packages within a terrane, such as a series of thrust sheets stacked one upon the other in what is commonly called a *duplex* configuration. Similarly, distinct lithologic packages that are separated by unconformities do not constitute different terranes. A volcanic sequence built on an ophiolite that is overlain by thick pillow basalt, red beds, shallow-marine and finally deep-marine pelagic strata, does not necessarily indicate two or more terranes. In fact this is the general vertical sequence displayed in the Wrangellia terrane, a terrane that may have originated in the western Pacific on the Gondwana margin but is now plastered along the edge of western North America from Oregon to Alaska. Thus, even though the definition of a terrane is quite precise, there is often room for competing interpretations when applying the definition to a given sequence of rock. In general it is better to err on the side of the splitters than in the camp of the lumpers; for once an assemblage is treated as a single entity, there will be a tendency to overlook data that would otherwise point out contrasting histories, of different elements, within the lumped terrane. If the starting point is with an excessive number of terranes, many future geologists are likely to derive pleasure from demonstrating a geological linkage among two or more of the preliminarily identified terranes, thereby reducing some unnecessary terms.

Size of terranes

Terranes may vary enormously in size, from subcontinental dimensions to bus-sized knockers. Within the mosaic of large continental blocks (terranes) that have fused during the past 250 m.y. to form Asia, many much smaller flakes and slivers represent fragments of seamounts, island arcs, and submarine fans. Information regarding these terranes is also important in order to ascertain the timing of accretion and the sense of movement within the Asian collage. The scale of the problem being solved generally dictates how finely one may wish to subdivide a region into terranes. For example, in making a North American continental compilation at a scale of 1:2 500 000 or smaller, the Franciscan complex of California cannot be subdivided practically into more than one or two terranes. At these scales the terrane is characterized as a collision or subduction complex that contrasts dramatically with the adjoining island-arc and continental terranes. However, at scales of 1:750 000 or larger, the Franciscan can be divided into 15 to 20 terranes – terranes that distinguish large accreted seamounts from submarine fan complexes, from high-grade blueschist, and from melange of a low metamorphic grade. Clearly, the processes that assembled these Franciscan terranes warrant an explanation more definitive than simple collision or subduction tectonics. Thus, the size of terranes is a condition of both the rocks and of the problem to be solved. The size question is tantamount to asking a stratigrapher, 'How thick is a formation?' The distributions of the grains of sand, the layers of strata, and the fault-bounded packages of rock do not chance, but how we read and interpret these distributions depends in part on our perspective and the types of the problems that are being solved.

Oceanic plateaus

Oceanic plateaus (Figure 2.8) are anomalously high parts of the seafloor that are not at present parts of continents or an active volcanic arc. Oceanic plateaus comprise fragments of continents, oceanic hot-spot tracks, remnant arcs, and other thick but poorly understood volcano–plutonic piles. Continental fragments such as Agulhas, Kerguelen, Exmouth Plateau, Lord Howe Rise, and Lomonosof Ridge occur relatively close to cratonal masses away from which they have rifted. But other possible continental fragments, such as Broken Ridge Plateau, are more dispersed; in this instance owing to a jump in the location of a spreading ridge and the subsequent rifting from the north margin of Kerguelen Plateau. Oceanic islands and seamounts such as the Mid-Pacific Mountains, and hotspot tracks, such as the Walrus, Hawaiian–Emperor, Louisville, and Ninetyeast Ridges, and the Canary Islands, are abundant features in the oceans.

Examples of remnant arcs include Bowers, Aves, and Palau–Kyushu Ridges. Several plateaus such as Ontong Java, Manihiki, and possibly the northwest part of Kerguelen are composed of oceanic island basalt and sediment near the surface but limited seismic reflection and refraction data suggest that they have a thick (30–40 km) continental-like structure at depth.

Oceanic plateaus are not tectonostratigraphic terranes. They are not fault-bounded crustal fragments; but they are destined to become terranes owing to the persistent movement and recycling of oceanic crust. When these thick and buoyant bodies encounter a subduction zone or transform margin, they will probably be accreted in some fashion to the adjoining plate, either an island arc or a continental margin. Examples of plateaus that have collided with subduction zones include the Umnak Plateau and Shirshov Ridge of the Bering Sea. These collisions may have contributed to the clogging of the Bering subduction zone such that subduction jumped southward to the present location of the Aleutian Trench, thereby trapping

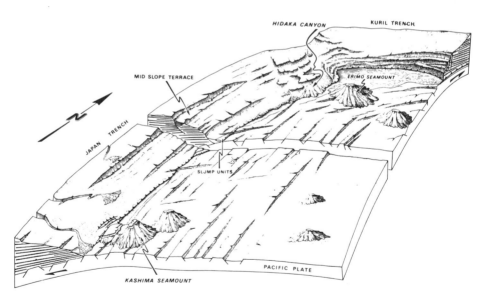

Figure 4.11 (a) Physiographic diagram showing the location of two seamounts (Kashima and Erimo) within the Japan Trench. This subduction zone displays superficial normal faults between the two collision zones; note also the extensional block faulting in the ocean crust just seaward of the trench. This diagram is constrained by detailed bathymetric data. (From Cadet *et al.* (1985))

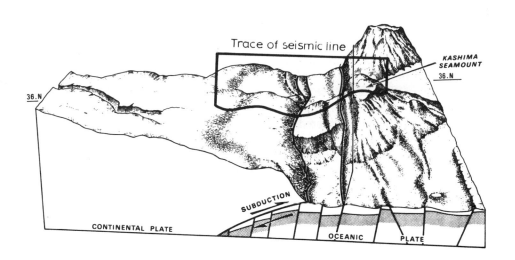

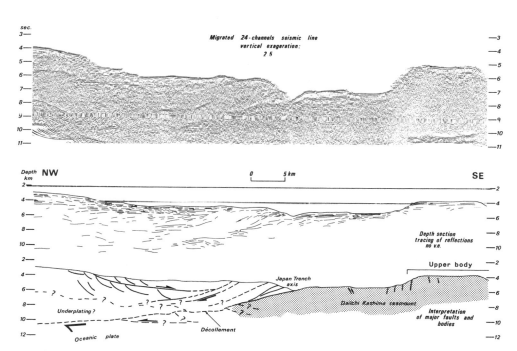

Figure 4.11 (b)   A closer view of Kashima Seamount locating the trace of a seismic line which shows the collapse of the frontal region of the seamount and the incipient subduction of a portion of it. Flat-lying faults below the principal decollement may be early signs of how the subducted seamount would underplate the accretionary prism arcward of the trench. (Data and interpretations from Cadet *et al.* (1987) and Lallemand, Cullota, and Von Huene (1989))

the Bering Sea plate and preserving in a pristine state an early phase of plateau–continent collision.

Examples of plateaus that are currently encountering modern subduction zones include the Nazca, Carnegie, and Louisville Ridges and the Ogasawara Plateau. French and Japanese geologists have recently collected seismic and detailed bathymetric data in the Japan Trench where seamounts are currently being subducted (Figure 4.11). In addition to collecting these data, they have made deep submersible dives to visit the seamounts in the trench setting. From this experience they have learned that basaltic seamounts upon entering a subduction zone partially splinter and collapse by normal faulting (Figure 4.11(b)). The smaller pieces tectonically descend beneath the leading edge of the arc system, the deformation front. The unresolved question is whether or not these pieces become underplated beneath the margin of the arc. Small bulges on the slope of the accretionary prism that have magnetic signatures suggest that seamounts lie in the subsurface, probably accreted by an underplating process within the framework of the prism. Such accretions may have caused the prism to rise, which helps explain the slumping and extensional faulting that is displayed in Figure 4.11(a).

Ontong Java is an enormous plateau, and when it encountered the Solomon volcanic arc its leading edge rode up and was obducted onto the arc. The basement rocks of several of the Solomon Islands are correlated with this plateau. An aspect of this collision is the rapid flip in the orientation of subduction, from southwestward along the northeast side of the arc to northeastward on the southwest side. When India crashed into the mass of Asia during the early Tertiary it both penetrated into and slid below the pliant continental domain of Asia, for a distance of at least 1500 km. The large mass of Ontong Java, upon colliding with the thin curtain of the Solomon arc, pushed the arc ahead and now carries it along as they override the new subduction zone that flipped to the southwest side of the arc.

Terrane maps

A terrane map is a derivative of a geologic map. The essence of a terrane map is the presentation of the packaging of terranes that compose the area being addressed. The principal data base is the stratigraphy of the region. The characteristics of terrane maps are still being experimented with. Some show only the inferred outlines of the principal terranes, for example the seminal 'suspect terrane map' of the Cordillera (Figure 4.12). Rarely, at present, is there any form of kinematic symbology along the terrane boundaries as the movement history of these boundaries is commonly complex, resulting from multiple episodes of amalgamation, accretion,

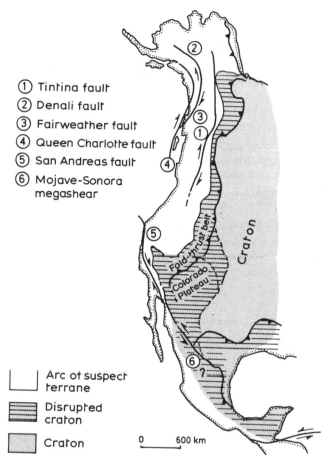

1 Tintina fault
2 Denali fault
3 Fairweather fault
4 Queen Charlotte fault
5 San Andreas fault
6 Mojave-Sonora megashear

Arc of suspect terrane

Disrupted craton

Craton

0   600 km

Figure 4.12 Area of Cordilleran suspect terranes, the portion of the craton that has been disrupted tectonically during the past 160 m.y. and the western limit of the craton. The important message conveyed by this map is the uncertainty in the geologic relations between the Precambrian crust of North America and the crustal elements that lie outboard of it. Stratigraphic and structural data indicate that this outboard area is composed of tectonostratigraphic terranes, and therefore the linkage of each terrane to the Precambrian crust is 'suspect', necessitating careful analysis to determine if the crust within each terrane is allochthonous, and, if so, by how much. (Modified from Coney (1981))

and dispersion. In order to record the kinematic history among the terranes, as well as the strain within them, one usually needs to incorporate study of an assortment of tectonic maps.

Some detailed terrane maps show, in addition to the terrane sutures, the distribution and limits of depositional overlap sequences or stitching plutons. The chronology of these features may be depicted directly on the map or in an accompanying tectonic assembly diagram. A critical feature of all terrane maps is the composition of the terranes, but rendering this on small-scale maps (less than 1:750 000) is generally not possible. Consequently, most terrane maps are provided with a test outlining a brief description of each terrane, or, alternatively, an accompanying sheet that contains stratigraphic columns depicting the salient stratigraphic relationships.

The basic data composing the ingredients of terrane maps are relatively objective. One may wish to attempt a characterization of the terranes in terms of tectonic kindred (e.g. island-arc, seamount, or continental affinities (Figure 4.13)) or label belts of terranes (for example Figure 4.14) or identify individual terranes (Figure 4.15) in the context of an accretion sequence. The efficacy of such attempts depends on both the extent of the data base and on the fundamental nature of the terranes. A terrane with a compound history does not lend itself to a simple tectonic kindred characterization; for example the stratigraphy of the Wrangellia terrane that crops out from Oregon to Alaska includes a history of island-arc volcanism, voluminous basaltic eruptions associated with a rifting event, subaerial to shallow-marine carbonate deposition, as well as abyssal marine conditions. The geologic contacts are all unconformities, so this is not a composite terrane. One may wish to refer to Wrangellia as a generic oceanic plateau that evinces multiple episodes of thermally driven uplifts and subsidences, typical of some oceanic settings, or one may wish to characterize Wrangellia on the basis of its island-arc basement, as has been done in Figure 4.13. Depending on the author's intent, any number of terrane characterizations are possible. The preliminary circum-Pacific terrane map was built with the motive of depicting terranes in terms of affinities to volcanic arcs, seamounts, sediment piles, or continental fragments. These divisions were further subdivided on the basis of (1) presence or absence of a continental signature, for example oceanic island arc versus Andean-like arc or pelagic deep-sea sediments versus continental-margin graywacke, and (2) whether or not their stratigraphy postdates the breakup of Pangea. The intent of this presentation is to facilitate a comparison between accreted terranes and modern plateaus and to distinguish the collage of terranes associated with the subduction of Panthalassa from terranes that record the recycling of older oceans.

Characterizing terranes on the basis of when they accreted to the continental margin is often difficult owing to extensive amounts of postaccretion dispersion. Most of the terranes in the Cordillera of North America accreted to the west margin of the continent between 160 and

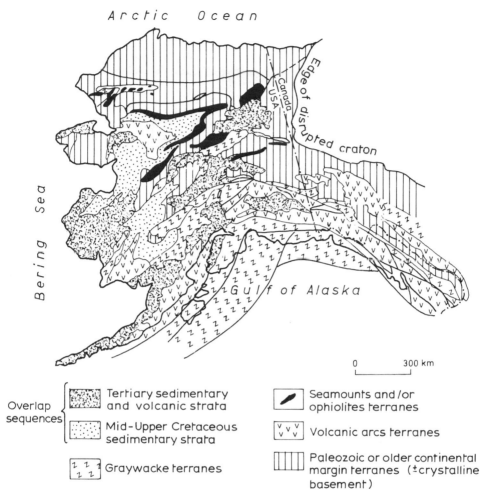

Figure 4.13 A more detailed terrane map than Figure 4.12 showing aspects of the terranes in Alaska classified in terms of tectonic kindred. From this map one realizes that the collage of terranes represents accretions involving a combination of oceanic plateaus of different types and slivers of continental crust probably dispersed from the North American continent. (Modified from Jones, Silberling, and Coney (1986))

60 m.y. ago but the post 60 Ma strike-slip faulting has amounted to hundreds or thousands of kilometers of spatial scrambling which both clouds the accretion history and in places relegates accretion to a second-order importance. The south margin of Eurasia, however, seems to

Figure 4.14 A very generalized rendering of the terranes and major oceanic plateaus of the circum-Pacific region. Orogenic belts represent distinct episodes of terrane accretion, and the ocean plateaus are classified in terms of tectonic kindred. They represent elements that will make up future terranes.

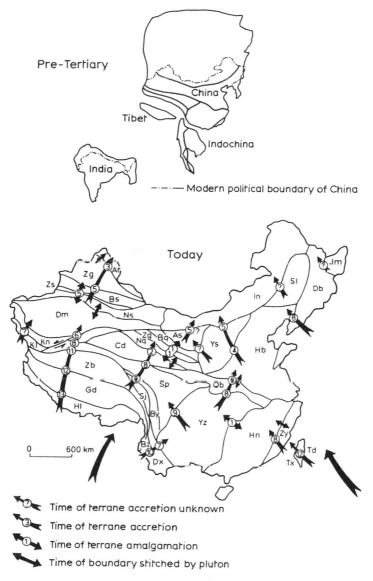

Figure 4.15 The tectonostratigraphic terranes of China with inferred ages that chronicle the timing of amalgamation and accretion. Note the systematic growth of China as a result of ferranes arriving at the margin of Eurasia from the southern oceans of the Tethyan realm. 1 = Silurian, 2 = Devonian, 3 = Early Carboniferous, 4 = Middle to Late Carboniferous, 5 = Permian, 6 = Late Permian, 7 = Early to Middle Triassic, 8 = Late Triassic, 9 = Late Triassic to Early Jurassic, 10 = Middle Jurassic, 11 = Cretaceous, 12 = Late Cretaceous, and 13 = early Tertiary. (Map from Ji and Coney (1985))

have preserved 200 Ma of southward outbuilding. This is well illustrated on the terrane map of Figure 4.15. Nonetheless, here, too, extensive postaccretion dispersion has occurred, both within the Tethyan orogenic belt as well as across the whole of the Asian continent, as a consequence of India impacting into the south margin of Eurasia (Figure 4.16).

The scale of terrane maps depends largely on the degree to which the compiler wishes to subdivide the crust into crustal fragments. The large

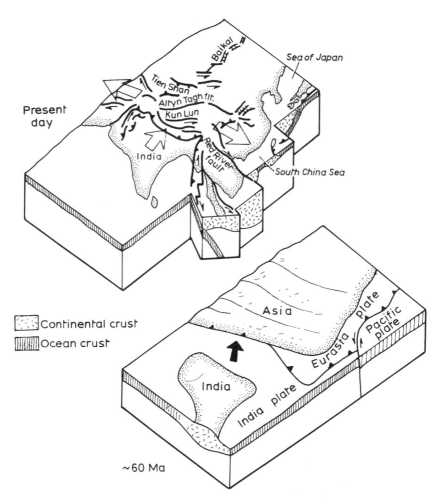

Figure 4.16 Block diagram depicting the continent of India colliding and indenting the Asia margin and the effect of eastward dispersion of continental crust. (Modified from Armijo *et al.* (1986))

fragments of island arcs or continental blocks in the upper 5 to 10 km of the crust are generally surrounded by zones principally composed of tectonized sedimentary rock, the clastic portion of foldbelts. These sedimentary terranes can be thought of in terms of a matrix to the collage of basement terranes. And this matrix may contain many small crustal fragments. The Franciscan terrane of California, Kurosegawa terrane of Japan, and the so-called ophiolitic terrane of the European Alps all contain discrete fault-bounded crustal fragments that constitute legitimate terranes when rendered on larger scale maps. At scales smaller than 1:5 000 000 the composite Franciscan terrane that consists of an assemblage of graywacke, basalt, and chert, locally with a high pressure and low temperature metamorphic signature, forms a discrete unit contrasting with the surrounding island arc and continental terranes. But at scales smaller than 1:750 000 one can see that the Franciscan is not an intractable melange but rather an orderly array of small terranes bounded by thrust and strike-slip faults admixed with overlap sequences, many of which have been tectonically kneaded into the terrane collage. These relations convey a complicated history of accretion and dispersion along the leading edge of a continental margin that is similar in style to the construction for the whole of the Cordillera.

4.4. PRECAMBRIAN TERRANES

The assortment of structural and stratigraphic characteristics that is normally used to differentiate terranes is commonly either missing or masked in Precambrian terranes, owing to regional metamorphism or erosion. Identifying Precambrian terranes, especially the older Early Proterozoic and Archean terranes, is something like asking an ornithologist to spot his favorite birds without their plumage. Nonetheless, such terranes must exist because plate-tectonic processes have been functioning for most, if not all, of Precambrian time.

Greenstone belts

In the most general terms, two basic assemblages of rock comprise greenstone belts. One involves a succession of volcanic rock, mostly basalt and andesite with minor amounts of the more felsic varieties such as dacite and rhyolite. These volcanic strata are depositionally overlain by a sedimentary interval of graywacke, shale, and quartzite. The 'maturity' of the sandstone increases upsection. The literature indicates that these types of greenstone belts are particularly abundant in North American Precambrian terranes. The second type of greenstone belt is a succession involving a bimodal suite of volcanic rock: ultramafic and basaltic lavas mixed with felsic tuffs and chert. These types of greenstone belts are reported principally from Africa and Australia. Neither succession is connected to its original basement.

At some localities a greenstone belt may have more than one succession, but the volcanic succession of the first type has never been seen to lie below the bimodal volcanic succession. Whereas greenstone belts were once thought to be intervals 10 to 20 km thick, recently acquired seismic-refraction data, supported by gravity modeling, now indicate that most are less than 10 km thick. Traditional interpretations have regarded greenstone belts as representing crustal sags or intracratonal rifts, but new interpretations include accreted island arcs or ophiolites that are structurally thickened above a basal decollement.

Are greenstone belts terranes? They get their name from the fact that they crop out in a linear pattern and because of the abundance of chlorite and other mafic minerals which imparts a green color to the rocks. Putative Archean ophiolites have been reported in the Wind River Mountains of the United States, the Slave Province of north-central Canada, and the Barberton Mountain Land of South Africa. These particular greenstone belts are certainly candidates to be evaluated in the context of composing allochthonous, fault-bounded crustal fragments. But not all greenstone belts are ophiolites. Thus, Precambrian geology is in a very exciting phase: much of the old dogma is being challenged, new interpretations are forthcoming, and field mapping and geochronology are focused toward gaining a fresh insight into the nature of Precambrian stratigraphy and tectonics. The distribution of Archean to Middle Proterozoic strata on a generalized Pangea reconstruction (Figure 4.17) aptly displays two of the conceptual options. Perhaps one is predisposed to argue that this distribution is *prima facie* evidence for the enduring presence of a supercontinent, and the outcrops of Archean strata are the iceberg-like tips of what in the subsurface is much more extensive. Alternatively, captured in this figure is an agglomeration of Archean terranes within a matrix of younger Proterozoic terranes. The continuity to the crust is not related to the configuration of crust as it formed, but is the simple consequence of accretion, the way bumper cars at a carnival occasionally form knots in a melee, halting the frenzied pace of revolving about the track floor.

Greenstone belts may provide the key to distinguish between the concepts of Precambrian supercontinents with steady-state volumes versus small Archean terranes and the continual growth of continents as a result of accretion. If greenstone belts indicate continental-margin settings, and if suture zones are housed within greenstone belts, numerous terranes should lie within these belts, just as the collision zones of Indonesia contain a collage of crustal fragments.

Canadian geologists and their colleagues have been particularly active in

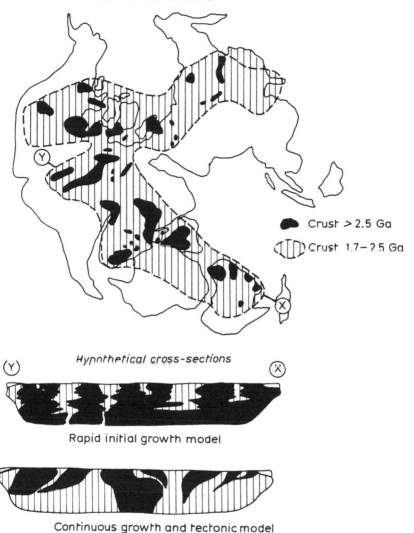

300 Ma reconstruction

● Crust > 2.5 Ga
⬭⬭⬭ Crust 1.7–2.5 Ga

Hypothetical cross-sections

(Y)   (X)

Rapid initial growth model

Continuous growth and tectonic model

Figure 4.17 Generalized map showing the modern outline of continents as if configured in a reconstruction of Pangea with the approximate outline of the known distribution of Archean and Early Proterozoic crust. The two schematic cross sections depict extreme viewpoints on the possible subcrustal distribution of Archean and Early Proterozoic rock; one suggests an early voluminous distribution of Archean crust with subsequent intrusions of minor amounts of Proterozoic crust, and the other indicates a more steady-state process of crustal formation with the framework of continents representing tectonic agglomerations.

reassessing the role of Precambrian plate tectonics in relation to the formation of the North American craton. A brief synopsis of some of these new ideas portends future thinking regarding the efficacy of Precambrian plate tectonics on other continents. We also need to know how to distinguish terranes when the rocks of deep crustal levels are exposed and

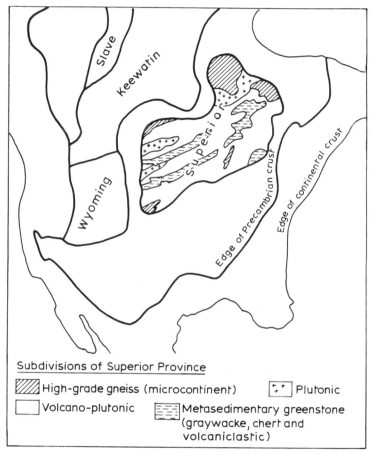

Subdivisions of Superior Province

High-grade gneiss (microcontinent)   Plutonic

Volcano-plutonic   Metasedimentary greenstone
(graywacke, chert and
volcaniclastic)

Figure 4.18 Outline of three crustal blocks older than 2 Ga within the craton of North America. Within the Archean Superior Province are shown the general outlines of greenstone belts. The internal structures of this area reflect thrust-nappe-style tectonics, and the overall composition and configuration of the greenstone belts suggest an agglomeration of volcanic island arcs, marginal-basin strata, and oceanic plateaus. The metaplutonic gneissic bodies (terranes) may be microcontinental blocks that are caught up in the collage. (Modified from Card (1986))

how to recognize and decipher agglomerations of terranes that have been overprinted by regional metamorphic processes.

Greenstone belts of the Superior province

A band of greenstone characterized by alternating sinuous belts of metavolcanic and metasedimentary strata lies between two high-grade gneiss terranes in the Archean Superior province of Canada (Figure 4.18). The high-grade metamorphic rocks are older (> 3 Ga) than the strata of the greenstone belts (2.6–3.0 Ga) and their structural fabric is not in alignment with the prominent east-trending grain of the greenstone belt. The greenstone metavolcanics include fault-bounded cycles of both types of volcanic successions. These high-level crustal strata have been intruded by voluminous plutonic rock, quartz diorite to granite in composition. The bands of metasedimentary strata consist of turbidites of poorly sorted volcaniclastic sandstone and conglomerate accompanied by some chemically precipitated ironstone and chert. Anatectic plutons locally intrude regions of high-grade metasedimentary strata.

Although other interpretations are possible, the data are consistent with the Superior greenstone belts as a succession of island arcs, marginal basins, and plateaus that define a broad collision zone between two continental blocks. The few instances of intermingled plutonic rock older than the greenstone volcanics may represent continental slivers (terranes) tectonically kneaded into the collision.

Aulacogens to sutures

One of the early attempts to apply plate-tectonic elements to the Precambrian involved a compilation of suspected rift features that may have accompanied the breakup of old continental cratons. Linear features trending at a high angle to an ancient cratonal outline may be the locale for thick accumulations of sedimentary and volcanic strata, and possibly mafic dike swarms. These are the characteristics of a failed rift arm, or aulacogen, formed when a hotspot fractures a plate into three segments (recall Figure 1.6). Modern examples include: the Benue Trough of west Africa, left behind when the two sister rifts propagated into the Atlantic Ocean; or the east Africa rift system, the failed arm that complements the ocean-forming rifts of the Red Sea and the Gulf of Aden.

Well accepted Precambrian aulacogens would seem to support the occurrence of Precambrian plate tectonics, and one of the first candidates was the 'Athapuscow aulacogen', the sequence of Proterozoic strata that lies along the southwest margin of the Slave Province in Canada. Although this interpretation helped support the notion of Precambrian plate tectonics, ironically, upon a re-evaluation of the Athapuscow aulacogen, it is now considered to represent continental transform faulting, associated with a continent–continent collision event, rather than the rifting of a single continent. High-precision zircon dating (see next section), gravity

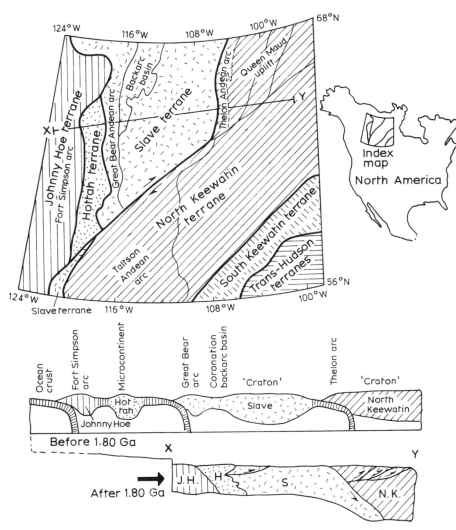

Figure 4.19 Map displaying the distribution of Archean and Early Proterozoic terranes in northwestern Canada with selected paleophysiographic elements superposed. The cross sections depict a generalized view of the tectonic conditions before and after the accretion of terranes that transpired during an approximate 180 m.y. orogenic period spanning the time from about 2.0 to 1.84 Ga. (Modified from Hoffman and Bowing (1984) and Hoffman (1987))

surveys, and regional-tectonic considerations favor this new hypothesis (Figure 4.19). The dating guides the sequencing of events. The gravity surveys display linear paired anomalies; this pattern has been interpreted to indicate a collisional suture marked by positive anomalies over the zone of accretion and a complementary crustal bulge stemming from the flexuring of the lithosphere. The regional tectonics include kinematically integrating the accretion of two continental blocks and a regime of transcurrent faulting that dies out in a zone of crustal thickening.

The Thelon magmatic zone (1.96–1.91 Ga) and the Taltson magmatic zone (1.99–1.92 Ga) are interpreted to be offset segments of an Andean magmatic arc (Figure 4.19). Right slip of 300 to 700 km along the Great Slave Lake shear zone delineates the collision between the Slave craton and the North Keewatin terrane (craton) of the Churchill province, yet at about the same time a host of subduction zones must have followed the trailing edge of the Slave craton. Here, arc volcanism is recorded from at least 1.94 to 1.86 Ga. An episode of possible backarc rifting is indicated by the Coronation Supergroup that is bracketed by 1.91 Ga rift-related basalt and the 1.88 anatectic Hepburn intrusives which indicate crustal thickening that followed the collapse of this inferred backarc basin. The Great Bear magmatic arc (1.88–1.86 Ga) and the Fort Simpson volcanic arc (1.86 Ga) indicate a paleogeography suggestive of the Indonesian archipelago. Between the Fort Simpson and Great Bear magmatic arcs lies a sliver of continental crust represented by the Hottah terrane. The assembly of the two cratonal blocks, the volcanic arcs, and exotic slivers of continental crust transpired during a period of 160 m.y. (2.03–1.86 Ga), a reasonable chronology even in the context of the slower plate motions of today. The presentation of this new interpretation is made particularly appealing by the analogy drawn with the modern tectonic setting of India crashing into Asia (Figure 4.16). The Slave terrane is analogous to India, the Queen Maud uplift similar to Tibet, and the Thelon fault relates to the north-trending strike-slip faults that traverse the Andaman Sea and coastal area of Burma east of India.

Terranes of cratons

Precambrian geologists have defined a number of provinces within the cratonal core of most continents. For example, I have already discussed how Figure 4.17 portrays the global distribution of Archean provinces which are generally surrounded by younger Precambrian orogenic belts. What are these crustal provinces, and what is the origin of the mobile belts that surround them? The North American craton provides some answers. Since 2 Ga, the North American craton has grown as a consequence of the accretion of island arcs and microcontinents, and the general shape of the

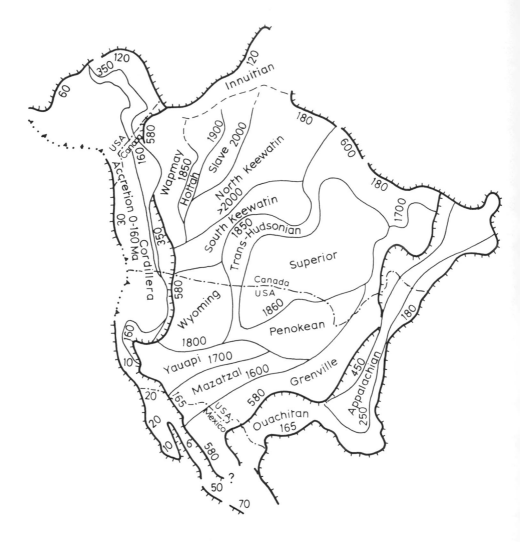

Dispersion scars, approx. ages for continent
Sutures, approx. ages accretion
Active accretion zones

Figure 4.20 Map of the North American continent, displaying the distribution of provinces and major orogenic belts; each presumably is composed of terranes. Also shown are approximate ages of accretion and the locations for the edge of the continent formed by dispersion processes at various times during the past 600 m.y. (Modified from Williams *et al.* (1989))

continent is the result of intervening episodes of rifting (Figure 4.20). The 2.6–3.0 Ga Superior greenstones and the 1.8–2.0 Ga Wopmay and 1.6–1.8 Ga Yavapai and Mazatzal orogens are prime candidates to be interpreted as regions that grew as a consequence of the accretion of volcanic island arcs. The Hottah, Slave, North Keewatin, South Keewatin, Superior, and Wyoming blocks are distinct microcontinental terranes that have been accreted to form the craton of North America. The Grenville province includes 1.1–1.4 Ga nonorogenic plutonic material that invaded the full expanse of the southern margin of this agglomeration of microcontinents, island arcs and ocean plateaus. The entire Grenville province underwent thermal metamorphism and crustal thickening about 1.0 Ga as a consequence of northwest-directed thrusting. Collision with a major continental block is suspected even though no remnant of such a block has been identifed as such. By the latest Proterozoic, about 600 Ma, the assemblage of microcontinents occupied the central area of a much larger supercontinent similar to the spatial relations of the craton of Antarctica to the supercontinent of Gondwana. A circumcontinental rifting event from 600 to 550 Ma sculpted the outline of the North American craton, creating much of the Precambrian outline as we know it today. Surrounding the cratonal core are still younger orogenic belts – Cordilleran–Innuitan–Caledonian/Appalachian–Ouachitan – that reflect a promulgation of accretion tectonics interspersed with rifting events. Thus, the tectonics of today provide an apt analog for the events of the past, and terrane analysis is appropriate for Archean, Proterozoic, and Phanerozoic provinces, from cratonal cores to modern mountain systems.

4.5 CONCLUSION

The many facets of plate tectonics help to provide exciting explanations for crustal-formation and tectonic processes. The model not only indicates how a variety of geologic settings are genetically linked – such as subduction zones–forearc basins–magmatic arcs–backarc basins – the model is also fully consistent with predictions regarding crustal dispersion and tectonic accretion. The geologic integrity of continental margins is suspect because of the nature of plate tectonics. With lithospheric plates moving at speeds averaging 5 cm/yr, and with orogenic belts often representing 200 m.y. of crustal convergences and translational processes, crustal displacements of 10 000 km are necessarily part of this history. Certainly not all, if any, of the crustal elements along a continental-margin orogenic belt have experienced kinematic displacements as great as this. Nonetheless, one should expect some degree of crustal dispersion and the rearranging of crustal elements, from an orderly configuration consistent with the tenets of plate tectonics to a more random pattern that also reflects plate tectonic processes. The collage of terranes that make up continental crust therefore reflects all aspects of its plate tectonic history. But can this history be reconstructed? The challenge to do such reconstructing is like having a puzzle box filled with only some of the pieces from more than one puzzle. The first task is to identify which pieces belong to which puzzle. Once that is done, efforts to reconstruct the various puzzles can be attempted, recognizing the limitations of having only a partial representation of each of the original puzzles.

5

Kinematics: measuring terrane displacements

5.1 OVERVIEW

The Kinematics of terranes generally refers to relative motions among terranes or between one or more terranes and a specified continent. In order to discuss terranes and the measurements of their relative displacements, several words warrant defining. *Allochthonous* is an adjective that denotes that a terrane, rock body, sand grain (or any other entity) has moved from its place of origin. The actual body that has moved is the allochthon, and the relative amount of movement is its allocthoniety. *Autochthonous* is an adjective that requires that the body (autochthon) in question has not moved from its place of origin with respect to the adjoining craton. *Parautochthonous*, as used here, implies relatively small displacements that are always in the domain of the continent within which the crustal fragment was formed (a parautochthon generally implies a fragment indigenous to the continent, with or without crystalline basement). *Exotic* implies that part of the history of a given terrane was in a foreign setting, spatially separate from the continent to which it is now attached. Thus, a collage of suspect terranes consists of allochthonous crustal fragments, but the allochthoneity of the individual terranes may vary considerably from one terrane to another, and not all terranes in a collage are necessarily exotic. For example, in the western Tethyan foldbelt involving the Pyrenees and European Alps, the terranes (nappes) are not likely to be exotic, because the ocean basins that lay between Africa and Europe were relatively small; whereas from about the modern Aegean Sea eastward, the ancient Tethyan Sea widened enough to afford the likelihood that many terranes in the eastern Tethyan foldbelt are truly exotic.

Geologists have numerous techniques available to measure crustal displacements. High-precision direct measurements of displacements between plates and across faults have documented short-term (hourly to yearly) small-scale displacements. For longer periods of time, on the order of 10^5 to 10^8 yr, analysis of magnetic lineations has revealed that plates have moved relative to one another from tens to thousands of kilometers, and the absolute motions among plates can also be reconstructed with reference to 'fixed' hotspots. Paleomagnetic studies have indicated travel paths for terranes, although the indicated displacements generally only provide the latitudinal or poleward component. Long-term APW paths graphically show the global wanderings of continents. Paleontologic data restricted to the past 580 m.y. may indicate large-scale displacements, but the resolution is poor, and these data generally only provide a north–south sense of movement. More classic studies of fault kinematics ordinarily do limit offsets in a temporal sense, but most reconstructions using these procedures have indicated displacements of about 500 km or less. The remainder of this chapter discusses in more detail the various methods used to ascertain the extent of allochthoneity of plates and terranes.

5.2 DIRECT MEASUREMENTS

Direct measurements

Radioastronomy and satellite laser ranging (SLR) have provided convincing evidence that lithospheric plates are moving in approximately the kinematic patterns prescribed by the plate-tectonic paradigm. The real-time measurements of the slowly moving plates are furnished by monitoring small changes in the distance between two or more fixed points on two or more plates. Using very long baseline interferometry (VLBI), large radio antennas record radio noise emanating from quasars in the outer reaches of the universe. Thus quasars are essentially stationary reference points because they are billions of light years from the Earth; any millisecond difference in the arrival time of the same signal can be converted to a distance measurement between the receiving stations ($V \times T = D$, where V is the speed of radio waves, T the millisecond time difference between stations, and D the distance between stations). SLR is similar to a high-precision surveying technique that utilizes satellites in the triangulations. Where these techniques have been applied in the Pacific (Figure 5.1), with stations in western North America, Alaska, the central and south Pacific, Australia, and Japan, calculations show that over the past several years Hawaii has been moving toward Japan at a rate of 87 ± 13 mm per year; this is essentially the rate estimated from ocean–floor magnetic lineation data by which the movement rate is recorded for the past several million years. Other such measurements indicate that Hawaii is moving toward Alaska at a rate of 52.3 ± 5.5 mm per year; which also is within the limits of error for the rate estimated from the magnetic lineation data. At present no techniques are available to measure rates over 10 to 10^5 yr

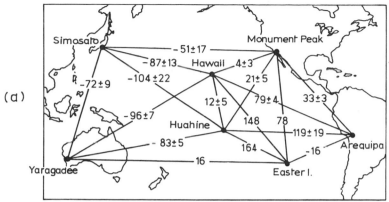

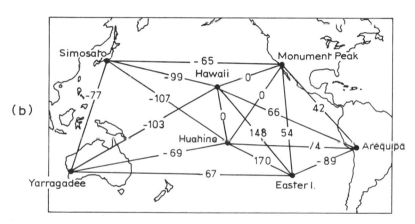

Figure 5.1 (a) Observed rates (mm/yr) of plate motion based on satellite laser ranging. (Data from NASA) (b) Rates (mm/yr) of plate motions between the geodetic stations based on paleomagnetic-lineation data and the plate tectonic model. The very short-term rates over the past several years agree remarkably with the inferred rates during the past several million years.

increments of time in order to determine if secular variations affect the rates at which plates are moving.

A new triangulation station is now being installed north of Tokyo, Japan, to determine if relative motion exists between the station and western North America; if both sites are on the North America plate no relative motion should occur between the two (cf. Figure 2.5). Other experiments are being conducted across the Atlantic between sites on the North America and Eurasia plates; preliminary results confirm the widening of this ocean as prescribed by plate-tectonic theory.

Similar experiments are being conducted in California in an attempt to monitor short-term movements along the San Andreas fault. In many other places, geodetic surveys are routine and are instrumented with equipment ranging from laser telemetric devices to high-precision theodolites in order to record short-term (hourly to yearly) crustal movements. The kinematic objectives include measuring the distortions of the crust in transpressional zones, the crustal inflation associated with volcanic activity, and the slip along fault planes. Numerous results confirm that the crust of the Earth is indeed a dynamic medium.

Present-day seismicity

The patterns of faults and folds in many modern orogenic zones may be so complex that they seem to defy any reasonable explanation. This complexity may reflect a nonrigid behavior for plates in a microplate setting and is more commonly an indication of superposed strain patterns resulting from reorientations of short-term stress within active orogenic zones. To understand the kinematics of these regions, one must first map the distribution of active faults (Figure 5.2), and first-motion studies must be performed to determine the relative sense of movement, as reflected by the propagation of seismic waves (Figure 5.3). These analyses not only indicate which faults are currently active but they also discriminate between normal, reverse, or strike-slip faulting (Figure 5.4). Although such studies are of no help in reconstructing ancient orogenic systems, a better understanding of the modern analogs instils an appreciation for the complexity of the tectonics that characterizes continental margins, something that is not so evident if one only applies the relatively simple rules governing rigid-plate dynamics of the plate-tectonic paradigm.

5.3 MAGNETIC LINEATIONS

For reasons that are not well understood, the polarity of the magnetic field periodically reverses. Approximately five to ten thousand years are required for the field to reverse. The reversal frequency is not cyclic. Reversals can occur as often as every fifteen thousand years, although the average frequency of reversals for the Cenozoic has been about one every 300 thousand years. For some intervals the magnetic polarity seemed to remain stable such as in the middle Cretaceous, when normal polarity persisted for 21 to 37 m.y. (depending on which published geomagnetic-polarity time scale is used). Since the Jurassic there have been about 93 reversals. The history of reversals is left indelibly inscribed on the oceanic crust (as discussed in Chapter 2), and thus enables us to reconstruct a network of relative plate motions for all surviving oceanic plates.

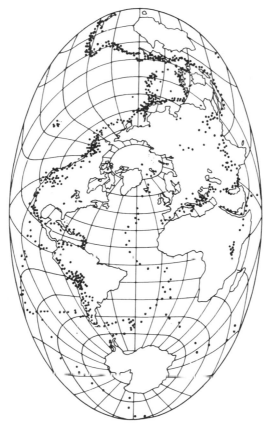

Figure 5.2 Generalized global distribution of major earthquakes since 1800. Most earthquakes are associated with modern plate boundaries or are reactivations of old plate boundaries.

A variety of procedures can be implemented to reconstruct plate motions from the magnetic-lineation data base. The essential information is the global map of magnetic lineations and the correlation of these lineations to a geomagnetic-polarity time scale. The orientations of the transform faults prescribe the relative direction of movement; transforms define small circles of rotation about the Euler pole of rotation. From the magnetic lineation data, plate motions can be inferred for intervals either between the successive reversal lineations or between the average ages for successive swaths of oceanic crust bounded by the reversal lineations. The actual configurations of magnetic lineation can be quite complex, and consequently the possibility of miscorrelating the ages of the lineations exists (Figure 5.5). With care, the history of plate motion can be reconstructed using the active spreading ridge as a zero baseline, as long as a

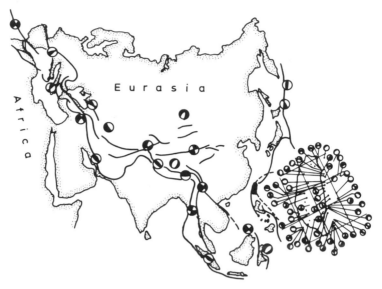

Figure 5.3 An arbitrary selection of earthquake focal mechanisms across southern Europe and Asia. An enlarged view of Taiwan is included to demonstrate the complexities of most regions of active faulting. Figure 5.4 is provided to assist in understanding how to interpret the 'beach balls'.

plate is not broken by a transverse fault. If two or more plates are joined only by spreading ridges, then the relative motions on any part of any such plate can be reconstructed and compared (Figure 5.6). Reconstructions of plate motions become problematic if previously intervening plates have disappeared as a result of subduction. Other problems include ridge jumps, nonuniform rates of spreading, and asymmetric spreading on opposite sides of a ridge.

At the present time the Pacific plate is sliding past the North America plate in the region of California. But how long has this been a transform margin? A technique of global circuitry, combined with the inference of symmetric spreading, offers a solution to the problem. The Pacific, Antarctic, Africa, and North America plates are separated only by spreading ridges; therefore, in the context of relative motion, if North America is fixed in space, it is possible to reconstruct the relative motions of these plates since about 90 Ma, because the oldest ocean crust that completely surrounds Antarctica has an age of 90 m.y. Knowing the positions of the spreading ridge of the Pacific plate relative to North America and assuming symmetrical spreading and the existence of one plate (Farallon plate) between this ridge and North America, one can find the relative motion of the now-vanished Farallon plate to the North

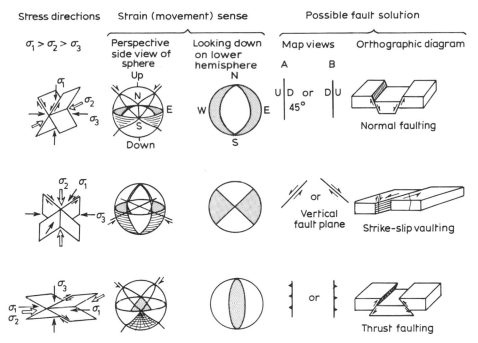

Figure 5.4 When the Earth ruptures, a host of shock waves radiate outward. Triangulating from three or more seismographs enables geophysicists to locate the region of faulting, and, from the character of the signal, determine the sense of movement along the fault plane. The traditional way of displaying the 'first motion' along faults is to use the lower-hemispheric projections of an idealized sphere centered around the rupture. The black area indicates compression and the white extension. Three end-member solutions are provided in addition to accompanying diagrams to aid in understanding stress directions and fault-plane solutions. Most faults are not purely strike-slip, reverse, or normal. When the motion is oblique the 'beach balls' take on a configuration that is often difficult to visualize, especially without practice.

America plate (Figure 5.6). Such reconstructions are necessary ingredients in the kinematic analysis of terranes and offer an explanation for the inclusion of oceanic terranes in the Cordillera, which presumably were carried by means of piggyback transport on the Farallon plate. Moreover, these analyses can be used to explain coastwise translational dispersion in the Cordillera during periods when the Farallon-plate motion had a northward component of motion relative to the North America plate.

Some terranes in the Cordillera, however, are believed to have experienced displacements that cannot be accommodated by the reconstructions

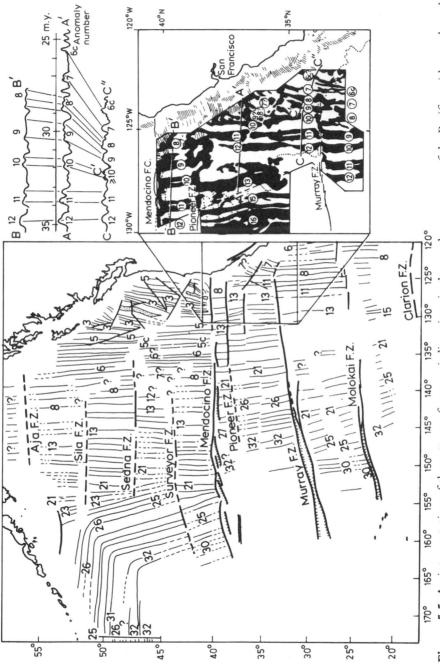

Figure 5.5 An interpretation of the pattern of magnetic lineations in the northeastern part of the Pacific, with an enlarged view of the area offshore from northern California illustrating the raw data. The anomaly numbers correspond to a nonlinear time scale that is constantly under revision; numbers 6 to 32 are for the approximate period 30 to 75 Ma. (Modified from Atwater (1970))

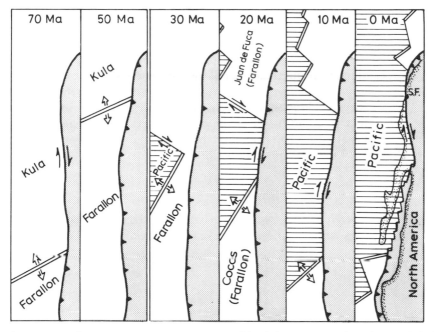

Figure 5.6 Plate reconstructions across an old subduction zone based on a global circuit linking the motions of the Pacific, Antarctic, Africa, and North America plates. Implicit in these reconstructions is the concept of symmetrical spreading and the existence of only the Farallon and Kula plates between the North America plate and the spreading ridge system that bounds the Pacific plate. The uncertainty in the location of the Kula–Farallon spreading ridge enables other kinematic scenarios along the west margin of North America before about 40 Ma to be made. (Illustration from Atwater (1970))

of the Farallon plate. Elements that could possibly introduce errors or misinterpretations in this reconstruction include the following:

(1) relative motion may have occurred between East and West Antarctica that would compromise the rigid-plate global-circuit analysis;

(2) other so-called mystery plates may have lain between the Farallon and the North America plates which also have vanished would have had their own motion relative to the North America plate; and

(3) the North Pacific contained a third plate called Kula, which possibly extended far to the south, such that prior to about 50 m.y. ago the region west of California was controlled by

Kula–North America motion rather than by Farallon–North America motion (Figures 5.7, 5.8). The evidence and implications of these alternative kinematic models will be discussed below; nonetheless, the rationale for plate geometries and trajectories and the rigorous treatment of the data just described are typical of many such inferences from throughout the Cordilleran and Tethyan foldbelts, and these uncertainties provide some of the fascination and much of the frustration that besets geologists in their attempts to rationalize the tectonic history of an area with plate-kinematic reconstructions.

Hotspot reference frame

The stability of a hotspot for tens to perhaps hundreds of millions of years is suggested by a straight line of successively active volcanic edifices that

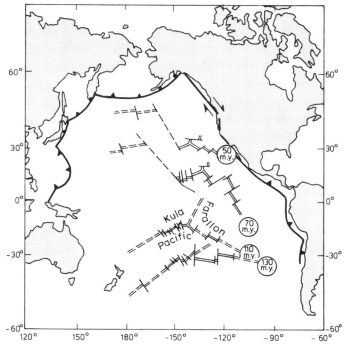

Figure 5.7 Inferred positions of the ridge–ridge–ridge triple junctions of the Pacific–Kula–Farrallon plates since 130 Ma relative to a fixed Eurasia. Reconstructions are based on the pattern of magnetic lineations shown on Figure 5.5. Note that despite the southward motion of the Pacific plate relative to the Farallon and Kula plates the Pacific plate too is moving north relative to Eurasia and North America. (Modified from Zonenshain, Kononov, and Savostin (1987))

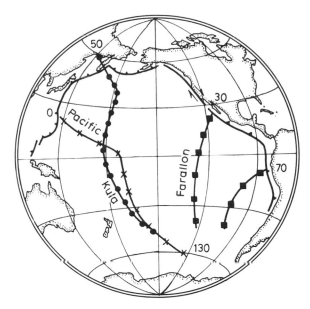

Figure 5.8 Based on the reconstructions of Figure 5.6 we can reconstruct the travel paths for distinct points on the Pacific, Kula, and Farallon plates since 130 Ma. For example, seamounts located on the Kula plate at 130 Ma at about 50° S and 180° W would have traveled 10 000 km northward and accreted along the Siberian margin by early Eocene; if the seamounts happened to have been located farther south on the Pacific plate, they would just now be reaching the Mariana Trench south of Japan.

record the trace of a plate moving over a hotspot. Other relationships would seem to indicate that hotspots are nearly stable with respect to one another. On the other hand, are the hotspots stable with reference to the Earth's spin axis? Seemingly not; preliminary analyses indicate that the hotspot reference frame wobbles a bit with respect to the spin axis. Nonetheless, the hotspot wobble is relatively small compared with the rate and extent of plate movement. Therefore, in addition to the global circuitry technique mentioned above, refinement of the interlinking of plate motions is made possible by calibrating these movements to the hotspot frame of reference.

The procedure entails mapping the oceanic crust of magnetic stages; these stages are the periods corresponding to the interval of time between successive magnetic reversals. The motion of the swath of crust for each stage is defined by an Euler pole, and the movement of coeval swaths from different oceanic plates can be compared via the hotspot reference frame.

Thus, the relative motion of plates can be reconstructed since about 130 Ma, the beginning of reliably identified hotspot tracks.

5.4 PALEOMAGNETISM

The basic principles of paleomagnetism were briefly discussed in Chapter 2 (see Figure 2.16). For kinematic studies of terranes, the primary objectives are generally to determine if any magnetic inclination anomalies indicate terrane movements either toward or away from the Earth's pole or if magnetic declination anomalies suggest crustal rotations, either for the terrane as a whole or for smaller blocks within the terrane. The following discussion summarizes how paleomagnetic data are incorporated into the kinematic analyses of terranes.

Sampling procedures

The stratigraphy and intraterrane-lithofacies relationships for any potential paleomagnetic site should be clearly understood. Stratigraphic information facilitates reasonable tectonic reconstructions. Furthermore, because corroborative studies are inevitably required to convince the doubting Thomases, an understanding of the stratigraphic framework is necessary in order to choose other sampling localities that will provide appropriate paleomagnetic results.

A variety of rock types can be sampled for a paleomagnetic analysis. Fine-grained volcanic strata are prime targets because they have potential for large magnetic signals and because establishing the paleohorizontal reference frame is generally easy. The magnetic moment of sedimentary strata is commonly more than five orders of magnitude less than that of volcanic rocks; nonetheless, with modern cryogenic magnetometers, even weak signals can be isolated to provide paleomagnetic inclination and declination information. As a rule, the ages of sedimentary rocks are known, and the orientation of the paleohorizontal can be determined with precision. Although nonmarine redbeds are a favored target, successful results have been obtained from subsea fan deposits (particularly mudstone layers including the fine-grained parts of turbidites), pelagic carbonate, fine-grained dolostone, and chert. Whether or not a given sequence of strata will provide a detectable and meaningful remanent magnetization cannot be known in advance.

Plutonic bodies may also produce reliable paleomagnetic signals. But even though plutonic bodies may have a strong remanent magnetization, not knowing the paleohorizontal with respect to the surface of the Earth is a problem. A means to overcome this problem is to select a batholith that crops out along strike for many kilometers. If the data are uniformly disparate from the expected results of an autochthonous body at the sampling site, and if no petrographic or metamorphic isograds trend perpendicular to the line of sample sites, then an inclination anomaly may appropriately be inferred, rather than post-emplacement tilting of the batholith. For example, if the sampling sites traverse a region 100 km across, and a 12° inclination anomaly is detected, tectonic tilting rather than latitudinal movement would be inferred only if numerous small tilted blocks were present or if the single block was tilted like a 'teeter–totter' such that one end rose 21 km relative to the other end. Presumably, careful mapping would detect either of these tectonic disturbances.

For all sites, regardless of the rock type, numerous samples must be taken. No hard and fast rules apply. A site may be a single bed or a thin sequence of beds. The importance of taking more than one sample per site is to provide statistical confidence in the reproducibility of the data, and multiple sites are necessary to average out secular variations in the orientation of the magnetic poles. If one is sampling volcanic rocks, where an instantaneous remanent magnetization is recorded, at least 15 flows should be collected in order to average out the effects of secular variations. For samples that record a post-depositional or chemical remanence fewer samples are required as each recording is a time average that commonly cancels the effects of secular variations. In these cases six samples per site, with at least six sites, should be adequate (but the more the better in order to derive a small standard deviation of the mathematical mean).

In addition to these sampling requirements, the meaning and accuracy of the data depend on a rigorous treatment that includes: (1) field orientation of both the cores and the host strata; (2) magnetization measurement that nearly always involves the removal of superposed magnetizations; and (3) field and laboratory tests to confirm an ancient magnetization. The two most common tests of ancient magnetization are

1. the *fold test*, whereby it can be shown that the dispersion of the magnetic poles is reduced after correcting for structural folding and therefore the age of magnetization is older than the structure; and
2. the *reversal test*, whereby magnetic core samples span episodes of both normal and reverse magnetic polarities. In such instances the magnetic signals are likely to be original magnetizations, for a strong superposed remanent magnetization would only have a single polarity.

Paleomagnetic results

The presentation of paleomagnetic data may be a mind-boggling affair for the uninitiated. Most presentations include plots showing the sequence of paleomagnetic azimuths as the sample is systematically 'demagnetized', that is, as the sample is subjected to incremental heating or to higher intensities of a randomly oriented, synthetic magnetic field. The critical element of reliable data is an invariant azimuth during the final steps of demagnetization. This reasoning assumes that the original magnetization recorded in the rock is the most resistant, and the persistent azimuth indicates that the superposed magnetic components have been 'cleaned' (removed) from the original magnetic orientation. The higher temperature and intensity demagnetizations degrade only the strength of the remanent magnetization, and do not alter the orientation. But understanding the diagrams that record this information can be a perplexing exercise in visualizing a convoluted trajectory through three-dimensional space. Difficulty in deciphering the demagnetization record and accepting the reliability of the final azimuth as the original magnetic orientation only adds to the mystique of the 'paleomagicians'.

Another important element in a paleomagnetic data set is the presentation of the azimuths of the remanent magnetizations, both before and after the correction for the tectonic tilt of the beds. These data may be displayed in a variety of ways. A common procedure is to show the intersection of the magnetic directions as unit vectors on the lower hemisphere of a stereographic diagram (Figure 5.9). The aspect of these data that is the most disturbing, when seen for the first time, is the large spread of the field of data points while the calculated mean with its error limits occupies a narrow range within the larger cluster. Thus, even though the raw data may be spread across a field of 30° or 40°, the calculated mean may only have an error factor of $\pm 2°$. How is this possible? It is simply an inherent property of the magnetic field and its secular variation caused by the movement of the pole about the Earth's spin axis. Remembering Figure 2.16(b), we observe that the locations of the magnetic poles in the northern hemisphere are spread throughout the Artic region; yet the mean of these data is precisely coincident with the pole of the spin axis. These types of data are analogous to daily recordings of temperature; in places like Chicago or Beijing the range of recordings may vary between 40° C in the summer and $-20°$ C in the winter, yet the average yearly temperature during the past century has not fluctuated more than 1° C. Accurately recording these kinds of data requires a large enough sample in order to define the mean statistically.

Another aspect that may cause some consternation is the 'hemispheric ambiguity', which is a consequence of not knowing the polarity of the magnetic field at the time of magnetic acquisition. In these instances the inclination data can be solved with either a southern or northern hemisphere solution; the correct answer depends on knowing whether the magnetic field had a normal or reverse polarity. Commonly the local

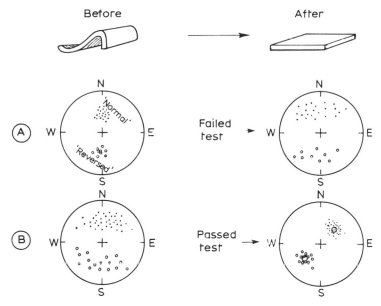

Figure 5.9 Hypothetical illustration of the behavior of magnetic data for a mid-latitude region where in case A the data fail the fold test, probably because of a strong overprint by the magnetic field during the past several million years, and in B magnetic data pass the fold test. The pass–fail grading scheme is based on the degree to which the data cluster about a mean before and after adjusting the magnetic orientations for the amount of tectonic tilt. A 'passed the fold test' signifies that the indicated magnetization was acquired before the structural folding event.

geology may indicate the most reasonable solution. The practice of selecting the least astonishing answer may persuade one to select the reconstruction involving the smallest amount of displacement; or in conjunction with fossil and other geologic data one could infer the more likely hemisphere.

Geologists are sometimes leery of accepting 'black-box' solutions, particularly when the results compromise their years of hard work or indicate displacements that had not been considered. Paleomagnetism, however, has proved to be a powerful tool that must be incorporated into every tectonic synthesis. The nature of magnetization and problems attendant on the collection of reliable data are such that all paleomagnetic results must be subjected to scrutiny. One's confidence in the results is always enhanced when conclusions from more than one laboratory agree. During the past decade numerous paleomagnetic studies have indicated crustal displacements and rotations that had previously escaped the geologist's eye. By incorporating these notions in a new look at regional tectonics, many workers have been convinced of the efficacy of the paleomagnetic conclusions. Even though skeptics still abound, an enlightened dialogue will surely direct the geologic community to acceptable and accurate solutions. What follows are a few paleomagnetic studies that exemplify the utility of this revolutionary tool.

Wrangellia

Disjunct segments of the Wrangellia terrane are embedded in the Cordilleran collage of terranes from Oregon to Alaska (Figure 5.10). One of the main characteristics of this terrane is a thick pile of Triassic basalt, which is an ideal medium for magnetic study. All of the principal fragments of this terrane have been sampled, and the results uniformly show inclination anomalies suggesting north–directed displacements from 10° to 17° N or S (Figure 5.11). The minimum amount of drift, assuming an origin in the northern hemisphere, is about 3000 km. Because of the

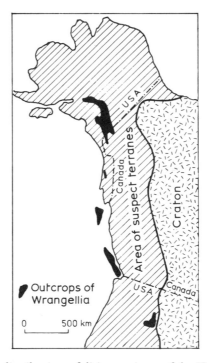

Figure 5.10 Present distribution of disjunct pieces of the Wrangellia terrane. See Figure 5.11 for a representation of paleomagnetic data from the piece of Wrangellia in southern Alaska.

uncertainty regarding the polarity of the Earth's field during the Triassic magnetization of the basalt, there remains the hemispheric ambiguity. The geometry of the data is such that the northern hemisphere option indicates 60° to 140° of clockwise rotation, while the southern hemisphere option indicates either 40° to 130° of counterclockwise rotations or 220° to 300° of clockwise rotations for the various terrane fragments. Both solutions have merit, and the correct choice is not known.

Magmatic arcs of western North America

Jurassic and Cretaceous magmatic arcs form one of the principal tectonic elements of the Cordillera of North America. Immediately following the

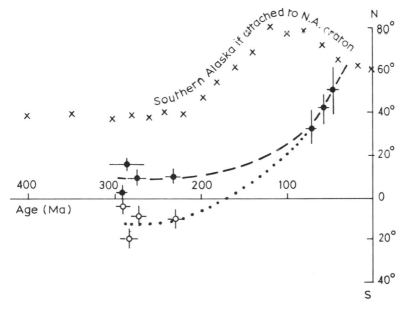

Figure 5.11 A plot of the latitude versus age for a selection of magnetic data from the Wrangellia terrane of southern Alaska. For the Triassic data, southern and northern hemispheric options are available, owing to uncertainty of the polarity of the magnetic field at the time of acquisition of the remanent magnetization. The theoretical latitude for the current position of Wrangellia is based on the APW path for North America; an Early Cretaceous accretion event would have required a greater northward travel path than is needed for the early Tertiary accretion owing to the southward movement of the Pacific part of the North America plate since the Late Cretaceous. These data do not constrain the east–west component of movement. (Modified from Hillhouse and Grommé (1984) and Panuska and Stone (1985))

formulation of the plate tectonic model, the positions of these magmatic edifices were incorporated into tectonic reconstructions that indicated the existence of a long, linear continental–margin subduction zone, analogous to the modern setting of the Andes of South America. This single subduction zone with its paired magmatic arc formed the basic framework of western North America. Where segments of this arc chain lie parallel to one another, it was assumed to indicate post-emplacement strike-slip faulting, locally for as much as 500 km, a distance which effectively doubled up the arc. However, all along the ancient arc from Alaska to Mexico, paleomagnetic studies involving volcanic and plutonic rocks and sediments that lie on the arc basement indicate numerous places where there is an inclination anomaly of between 10° and 20° and clockwise-rotated declination anomalies of 60° or more. These paleomagnetic data combined with geologic studies suggest that 1000 to 2000 km of northward transport took place along the margin of North America during latest Cretaceous to earliest Tertiary time; only the Peninsular Ranges batholith of southern and Baja California experienced such long-range displacement in post-Eocene time (Figure 5.12). No one in the geologic community predicted translations of such magnitudes. Nonetheless, the general geodynamic setting is not much different from an Andean arc, as was previously inferred, except that the various arc pieces seem to have been tectonically shuffled much more than expected.

India and Tibet

A host of geologic, paleontologic, and geophysical characteristics indicate the allochthonous aspect of the different terranes within the Tethyan foldbelt. Published data confirm the northward movement of India and Tibet from a location along the northern margin of Gondwana. Figure 5.13 is a cartoon displaying the results of a paleomagnetic study that constrains the time and amount of the displacements. The questionable aspect of this rendering of the data is the seemingly parochial persuasion that what is now together must have always been together. This reconstruction exemplifies a fundamental problem inherent with paleomagnetic data. Because the Tethyan foldbelt is oriented largely east–west, longitudinal movements along the strike of the foldbelt cannot be differentiated.

Another study involving terranes in Tibet indicates that they have been displaced cratonward approximately 1500 km since accreting to the margin of Eurasia. This amount of movement corresponds to the inferred crustal shortening of the area that consequently resulted in the enormous crustal thickness beneath the Tibet plateau as well as the eastward

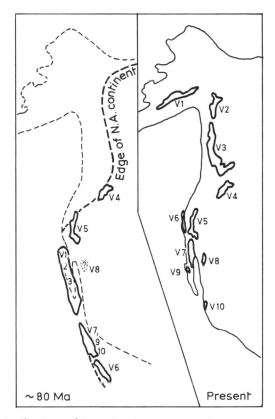

Figure 5.12 Distribution of Jurassic–Cretaceous volcanic arcs in their present configuration and a reconstruction for about 120 Ma based on a large volume of published paleomagnetic data.

expulsion of crustal fragments from within Eurasia which made room for the northward indentation of the Indian craton (cf. Figure 4.16).

New Zealand

The South Island of New Zealand is composed of several terranes involving Paleozoic continental strata, and Mesozoic volcanic island arcs consisting of thick accumulations of continental-margin graywacke, and oceanic material (Figure 5.14). Geologic studies have indicated that these terranes did not amalgamate until the middle of the Cretaceous; nonetheless, paleomagnetic studies involving Triassic and Jurassic strata reveal

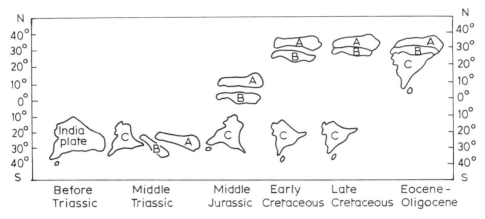

Figure 5.13 Latitudinal and rotational configurations for India and Tibet based solely on paleomagnetic data. Because of the complicated history of accretion tectonics since the latest Paleozoic along the south margin of Asia, terranes A, B, and C as viewed after the Oligocene are unlikely to have been in the same configuration before the Triassic. Longitudinal movement is not constrained by paleomagnetic inclination and declination data; therefore, east–west dispersion of terranes within the Tethyan foldbelt is poorly constrained. (From Zhu and Teng (1984))

magnetic inclinations suggesting deposition at a latitude of approximately 66° S, which is roughly coincident with the latitude of a segment of the margin of Gondwana where New Zealand would fit by simply palinspastically closing the Tasman Sea and Indian–Pacific Ocean south of it. Thus, these relations would seem to indicate that the only movement of these terranes is the latest Cretaceous to middle Tertiary dispersion coincident with the opening of new oceanic domains. Are the geologic interpretations necessitating terrane accretions faulty; or, alternatively, did all pre-amalgamation displacements involve east-to-west translations? The amount of ocean in the southern hemisphere in Gondawana time was extensive (Figure 5.15). Quite possibly New Zealand is analogous to the Cordillera of North America, where the terranes along the continental margin have been shuffled owing to northward translations; in the case of New Zealand, however, the translations may have involved translational slip along the continental margin of Gondwana or rifting and transoceanic dispersal from one part of Gondwana to another. Additional paleomagnetic data, sampling a variety of time-stratigraphic horizons, may help to determine which reconstructions are correct.

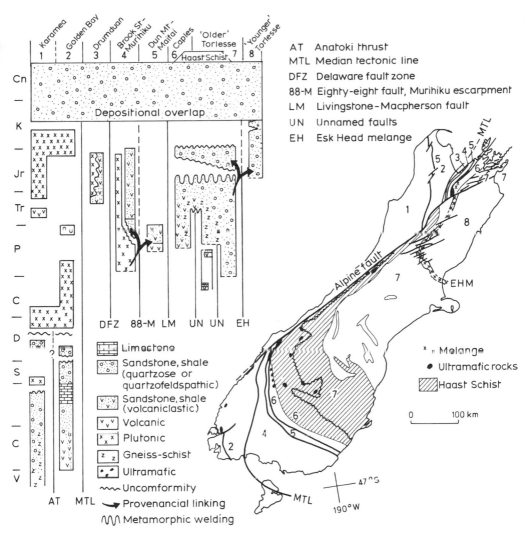

Figure 5.14 Terranes and a tectonic assembly diagram, South Island, New Zealand. The Dun Mountain–Maitai terrane forms a thin, nearly vertical sheet in the region of the Alpine fault. The linear map of this terrane intersects the fault at a high angle, and this, combined with the aforementioned features, makes effective piercing points indicating ca 460 cm of right slip. Another ca 100 km of right slip is recorded by offsets along the trend of the Esk Head Melange. (Illustration modified from Bishop, Bradshaw, and Landis (1985))

Figure 5.15 An early Mesozoic view of Panthalassa showing the extent of open ocean at 66° S, the Jurassic paleomagnetic latitude determined for the Brook Street–Murihiku terrane of New Zealand. Thus, paleomagnetic data do not constrain whether or not the sites of deposition of the terranes of New Zealand were along the continental margin of Gondwana or in an intraoceanic setting, as is indicated by the petrology of the strata. (Illustration modified from Zonenshain *et al.* (1989))

Franciscan seamounts

The Central Belt terrane of northern California is principally a broad melange in which a variety of exotic blocks are immersed in a mudstone matrix. Included in this assemblage are blocks of 101 to 88 Ma limestone that represent the carapace to seamounts; in most instances, the limestone caps have been tectonically severed from their volcanic basement. These pieces of limestone range in size from boulders to blocks an acre or more across. The random orientations graphically demonstrate somersault tectonics resulting in the melange fabric. Nonetheless, these blocks have been the object of noteworthy paleomagnetic studies. Bedding characteristics and high-precision dating of planktonic forams has enabled geologists to determine the original depositional orientation of each limestone block.

Nearly half of the blocks are inverted to varying degrees. When the assemblage of paleomagnetic data is rotated in order to place each block into its original horizontal orientation, the paleomagnetic poles form a tight cluster. This extraordinary foldtest demonstrates the authenticity of the remanent magnetization. The age of the strata correlates with the long Cretaceous period of normal polarity, and, if one assumes that the age of magnetization corresponds to the depositional age, the data have no hemispheric ambiguity. The results indicate that these limestones were deposited approximately 17° south of the equator. Regional overlap sequences require that accretion occurred sometime between the Late Cretaceous (post-90 Ma) and the middle Eocene (pre-50 Ma). Thus, the transit from south of the equator to northern California, a distance of no less than 6000 km over an interval of time no more than 38 m.y., suggests plate velocities of at least 14 cm/yr; the poorly constrained time of accretion allows that the velocity could have been nearly three times as fast. From Figure 5.16, however, one can imagine a kinematic scenario where the seamounts formed on the eastern part of the Kula plate and rode with it to the north. If generated on the Kula plate the seamounts would have arrived in the California area at about 50 Ma; from the range of possibilities this reconstruction favors the slower velocities.

Rotations

In most paleomagnetic studies where an inclination anomaly is determined, the data also indicate varying amounts of block rotation, usually in a clockwise sense if associated with right-slip transcurrent faulting and counterclockwise in instances of left-slip transcurrent faulting. In several studies, however, paleomagnetic declination anomalies have been determined in the absence of any inclination anomalies. Rotated blocks can be of any size. In central California blocks as small as volcanic necks that happen to lie within a shear zone have spun like small tops. The Iberian Peninsula rotated counterclockwise as it slid to the east away from Brittany creating the Bay of Biscay in its wake. In effect, this is a small-scale version of the opening of a large ocean basin, where entire continents rotate away from the spreading ridge. In other instances, flaps of crust many hundred kilometers long and less than half as wide have seemingly pivoted like the hands of a clock, such as the Transverse Ranges of California, the Apennines of Italy, or the Brooks Range of Alaska. Entire foldbelts have been bent into oroclines, such as all of southern Alaska and the Bolivian Andes. To accommodate these rotations a decollement must lie either within the crust or deeper in the lithosphere across which the upper plate can slide and rotate, 'frisbee' style.

Thus, the experience from these studies indicates the importance of

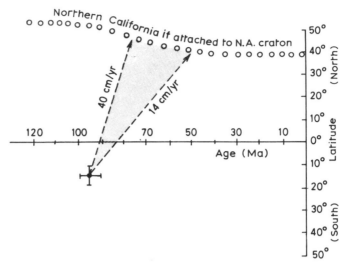

Figure 5.16 Latitude versus age diagram for the Laytonville Limestone (exotic knockers imbedded in the Central Belt terrane, the large melange terrane of the Franciscan assemblage of northern California). The southern–hemisphere site of deposition was established on the basis of paleomagnetism. Note error bars reflect degree of precision for both paleontologic age and magnetic inclination data. Stratigraphic data require that these limestone fragments were accreted to California after 90 Ma and before 50 Ma. Minimum terrane movement rates of 14 to 40 cm/yr are thus indicated. This is a minimum range because these paleomagnetic data only provide information regarding the polarward component in motion. Combining this information with the reconstructions shown on Figure 5.6 suggests that travel must have been on the Kula plate, and therefore a travel rate of about 16 cm/yr is implied. (Data from Alvarez *et al.* (1980) and Tarduno *et al.* (1986))

incorporating paleomagnetic investigations in all regional tectonic syntheses. The paleomagnetic signature of a terrane is every bit as important and as revealing as its fossils or lithologic characteristics. Some geophysicists have suggested that because geologic data are often equivocal our confidence in tectonic reconstructions cannot be secured until research areas have been blanketed with paleomagnetic studies. There have even been allegations of 'secret meetings' among paleomagneticians where geologists were excluded for fear that their conservative tendencies would compromise the essence of the paleomagnetic data. Such meetings, however, are probably events of the past – for the revolution is behind us.

5.5 PALEONTOLOGY

Principles

The past and present distribution of plants and animals is a consequence of evolution and the effects of environmental pressures. The ecologic niches occupied by species are defined in terms of climatic factors: temperature, light, chemistry, pressure (atmospheric and hydrostatic). Individuals may freely occupy any part of their niche, though environmental barriers and competition with other species further restrict the distribution of species. The tolerance to climatic conditions varies among species and therefore niche sizes among groups of species can be quite different. Some species are particularly cosmopolitan (humans, rats, and houseflies) while others are more geographically restricted (sea cucumbers, corals, and caribou). More benign climatic settings host a greater diversity of species than hostile settings, even though the total biomass may be roughly equivalent (more species of mosquitoes exist in the tropics than in the Arctic, but the fewer species in the Arctic abound with individuals).

The foregoing are oversimplifications of ecologic and biogeographic principles. Nonetheless, they provide a basis for understanding a variety of paleontologic observations, some of which support continental drift and displacement history for terranes in an orogenic collage.

Wallace's line

About the time Charles Darwin was formulating his theory for the origin of species Alfred Russel Wallace was recording enigmatic differences within the population of plants and animals of the Indonesian archipelago. In an 1858 letter, he wrote: 'in the Archipelago there are two distinct faunas rigidly circumscribed, which differ more than those of South America and Africa, and more than those of Europe and North America. Yet there is nothing on the map or on the face of the islands to mark their limits. The boundary line often passes between islands closer than others in the same group. I believe the western part to be a separated portion of continental Asia, the eastern the fragmentary prolongation of a former Pacific continent.' By believing that paleogeographic factors control the modern faunistic differences between some islands, rather than any presently existing environmental barriers, Wallace anticipated modern plate tectonics.

Lungless salamanders

The family Plethodontidae are primitive salamanders that as a group are most unadventurous. Individuals ordinarily do not stray more than a few meters from the place where they were born. A variety of distinct species occupying similar habitats exist from southern California to central America. Species are limited to particular tectonostratigraphic terranes. An intriguing hypothesis, but probably not the only possible explanation for their zoogeographic distribution, states that these species evolved from a common stock sometime during the Late Cretaceous in the region now occupied by Central America. The northward dispersion of terranes since that time isolated different communities of salamanders and they have subsequently evolved into terrane-specific species. These animals are terrane riders rather than island hoppers.

Paleozoic floras

Late Paleozoic floras show a worldwide provinciality. Two cold-climate floras are represented by the Gondwana (*Glossopteris*) flora for the southern hemisphere and the Angora flora for the northern hemisphere. Three warm-climate floras (Cathaysian, North American, and Euramerican) occupied different low-latitude regions. When the fossil localities for these different warm-climate floras are plotted on the reconstruction of Pangea, it is evident that prior to the Triassic the area now occupied by southern Asia must have been detached and located further south in the tropical Tethyan Ocean (Figure 5.17). The 'mixed' Cathaysian–*Glossopteris* assemblage in Turkey, Iraq, and New Guinea leaves open the question as to whether or not these allochthonous parts of Asia were part of the north margin of Gondwanaland or insular continental landmasses in the southern Tethys.

Fusulinids

Fusulinids are large Foraminifera roughly the size and shape of small raisins that became extinct by the end of the Paleozoic. During the Permian there were two major families, the Schwagerinidae and the Verbeekinidae. The Verbeekinidae are known from throughout the circum-Pacific region with the highest density in the equatorial region of the Paleotethys; there they occur both in autochthonous continental strata and in allochthonous terranes. In high latitudes their occurrence is restricted to regions of suspect terranes (Figure 5.18). At similar latitudes the Schwagerinidae characterize the fusulinid populations of the autochthonous continental margins. Thus, the distribution of the so-called Tethyan fusulinids seems to demonstrate the dispersal to high latitudes of late Paleozoic equatorial terranes accompanying the subduction of Panthalassic and Tethyan oceans.

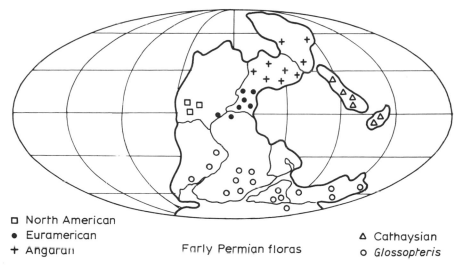

□ North American
● Euramerican
+ Angaran   Early Permian floras   △ Cathaysian
                                                     o *Glossopteris*

Figure 5.17 Late Paleozoic reconstruction of Pangea with the generalized distribution of five distinct floral assemblages. (Illustration modified from Stauffer (1985))

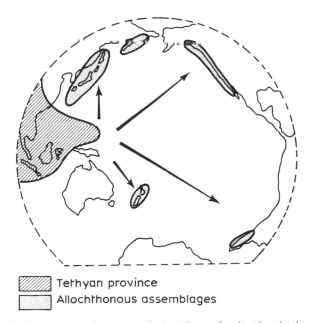

▨ Tethyan province
☐ Allochthonous assemblages

Figure 5.18 Distribution of the so-called Tethyan fusulinids which suggests that the high-latitude terranes containing these Late Paleozoic assemblages originated in an equatorial setting. (Illustration modified from Ozawa and Kanmera (1984))

Monotis

Monotis is a family of Late Triassic scallop-like mollusks that occur in the post-Pangea accretionary zones of the circum-Pacific region and Tethyan foldbelts. Because these thin-shelled individuals lived attached to floating marine algae, their distribution may reflect zoogeographic provincialism, postmortem dispersal by ocean currents, or post-depositional terrane displacements. All of the *Monotis* localities in the western Pacific lie within regions of suspect terranes; consequently, distinguishing between means of dispersal is difficult. In North America, however, three *Monotis* faunas lie within the cratonal realm, and their distributions seem to reflect a paleobiographic zonation corresponding to low, middle, and high latitudes (Figure 5.19). The putative low-latitude fauna – assemblage 'B' – is also widely distributed in more northerly latitudes of the Cordilleran area of suspect terranes. This distribution probably reflects the northward dispersion of terranes bearing *Monotis* of assemblage B.

The aforementioned examples were chosen to show how the distribution of both modern and fossil floras and faunas may reflect large-scale displacements. Because many factors control the limits to biogeographic provinces, fossil data alone may not provide a convincing argument for, or a unique solution of, terrane displacements. But in concert with other data, paleontologic information can help constrain the displacement history for a region of suspect terranes.

5.6 FAULT-PLANE SOLUTIONS

Since the 1950s, geologists have proposed offsets of as much as 500 kilometers along strike-slip faults, and Alpine geologists for the last century have been providing evidence for large-scale crustal shortening in foldbelts. These kinematic studies commonly involve intraterrane structures or final phases of terrane displacements associated with post-accretion crustal consolidation or continental-margin dispersion tectonics.

Transcurrent faults

Plate tectonics predicts large amounts of lateral crustal offset, and the concept of transform faulting explains the conundrum of how to terminate strike-slip faults that exhibit large displacements; namely, the crustal motion is taken up by or transformed into either a subduction zone or a spreading ridge. Measuring the amount of slip (displacement) along such faults is achieved by identifying the two offset segments of a unique geologic feature. The broken element should be a narrow feature that intersects that fault at a high angle. Magnetic lineations on ocean crust are

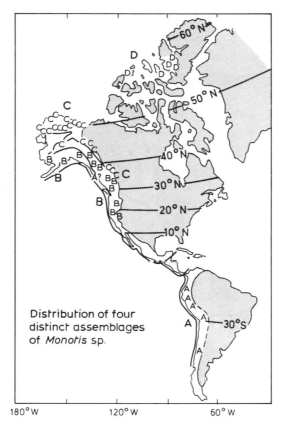

Figure 5.19 The present location of North and South America superposed on inferred Triassic paleolatitudes. The distribution of four assemblages of *Monotis* species are shown as A, B, C, and D. Regions of suspect terranes are unpatterned. Representatives of assemblages B, C, and D crop out on autochthonous North America, indicating a north-to-south zonation. Where assemblages of B and C occur within suspect terranes, they seem to be shifted northward, as demonstrated particularly by B. The original site of deposition of assemblage A is not known, and the one occurrence of A in the Alexander terrane, southeast Alaska, is particularly anomalous. (Illustration from Silberling (1985))

ideal for this purpose. For land-based geologists, paleoshorelines or lithologic pinch-outs may provide suitable lines of intersections, often called piercing points. The sense of displacement of offset planes, however, can be ambiguous; for example, normal or reverse faulting will create a map pattern that mimics a pattern of offset by lateral slip (Figure 5.20). Lateral slip is also possible such that geologic planes, such as bedding planes or sills are not offset. For these reasons purported offsets of metamorphic

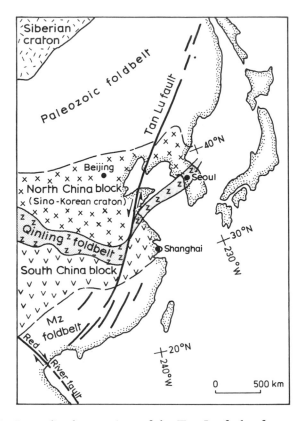

Figure 5.20 Generalized map view of the Tan Lu fault of eastern Asia. Many different offsets have been proposed for this fault including left slip of as much as 1500 km. In this rendering, the Qinling foldbelt and its inferred continuation to the east suggest at least 500 km of left slip, whereas the Proterozoic block of north and south China seems to show no displacement. Other enigmas are how the trace of the Tan Lu fault dies out to the north and south while being such a prominent feature for nearly 4000 km, and how the more recent right slip relates to the large-scale left-slip motion.

isograds or stratigraphic horizons should be viewed with caution. Vertical or near-vertical planes that intersect a vertical strike-slip fault, however, provide an appropriate means for measuring the component of lateral slip; for example igneous and sedimentary dikes or vertically disposed suture zones can be useful features for such measurements. An estimate of slip is also possible using bulbous rock bodies – such as plutons or volcanic edifices – if the amount of offset is much greater than the maximum dimension of the unit. The problem with this kind of measurement is knowing with certainty that the correlated features are in fact offset pairs. Studies involving the Alpine fault of New Zealand and the San Andreas fault of California are useful in illustrating these points.

The Alpine fault system, which forms part of the juncture between the Pacific and India–Australia plates (Figures 4.2, 5.14), is a transform fault between the Hikurangi Trough and the Puysegur Trench. Plate-kinematic studies predict approximately 1000 km of offset between the two lithospheric plates in New Zealand. Because of the orientation of the Alpine fault, a significant component of compression along the fault trace has produced as much as 25 km of vertical uplift. The erosion that accompanied this uplift has eliminated most lithologic features that could have made piercing points, but an ophiolite-bearing suture between the Caples and Hokonui terranes of Figure 5.14 clearly demonstrates approximately 460 km of right slip. Furthermore, the right-stepping offsets of the Esk Head melange in the north part of South Island record an additional 50 to 100 km of offset along subsidiary faults in this transform zone. Because the aggregate amount of displacement is still ca 500 km less than that predicted by the plate-tectonic kinematic reconstructions, extra displacement is inferred along offshore faults as well as by imperceptible shear distributed across a broad region of the plate boundary.

The San Andreas fault is another transform boundary. In this case, the faults links the Mendocino transform with the East Pacific spreading ridge in the Gulf of California (compare Figures 2.7 and 4.12). In this instance essentially no differential vertical displacement occurs along the 1000 km trace of the fault. A multitude of geologic piercing points have been identified that indicate approximately 305 km of right slip since the Middle Miocene (about 12 m.y. ago). These features include lithofacies pinch outs, ancient shorelines, basin troughs, and a volcano. Older features such as Eocene submarine-fan channels and distinctive Precambrian stratigraphic units are offset by similar amounts, which is consistent with the post-30 Ma displacements predicted by plate-tectonic kinematic studies. In central and northern California, however, Cretaceous batholithic rocks are seemingly offset by 510 km and this has invoked the idea of a pre-Eocene 'proto-San Andreas'. Because Precambrian strata in southern California do not indicate this extra 205 km of displacement, the missing proto-San Andreas fault in that area presents an enigma. Paleomagnetic data may have provided the solution. Inclination anomalies from Upper Cretaceous strata as well as from the inferred displaced batholithic bodies indicate not 500 km of northward movement, but at least 2000 km of displacement. Thus, the inferred ties of batholithic rocks across the fault are in error, for they are not matching pairs. Instead, the pre-Eocene strata seem to be exotic to

California and the post–30–Ma displacement along the San Andreas fault is only the most recent phase of a longer history of continental-margin dispersion.

The Tan Lu fault system (or Tancheng–Lujiang) forms a 4000 km long trace that trends N30° E through eastern China, Korea, and Russia. The long vertical fault plane is suggestive of a strike-slip fault. Kinematic features along the fault indicate a complex history, involving an older episode of left-slip superseded by younger right-slip. At the north, the trace is lost in the Amur River valley. Its southern end splays into several small faults (Figure 5.20). Regional geologic features cut by the fault seem to indicate anywhere from zero to 1500 km of left-lateral displacement. If the larger estimates of offset are correct, how the motion is accommodated at the southern terminus remains enigmatic.

The older history of the Tan Lu seems to be a mirror image of the right-lateral faults of western North America, both systems reflecting shear coincident with long-term northward movement of oceanic plates in the Pacific basin. The absence of numerous unequivocal features indicative of slip along the Tan Lu may be due to differential vertical displacements, concomitant with erosion, which has erased the matching pairs and disposed the cratonic blocks in a way that seems to show no slip.

Strike-slip faults that traverse a collage of terranes, for example the Denali and Kaltag faults of Alaska and the Anatolian fault of Turkey, are difficult to analyze for total displacement, because one is never certain about the original configuration of geologic elements on either side of the fault. Does the appearance of offset bodies reflect post-accretion dispersion or a geometry indicating pre-accretion fragmentation?

Most major strike-slip faults have similar controversies surrounding interpretations of displacement. The fundamental lesson to be learned is the importance of distinguishing between unique solutions of slip involving geologic lines that pierce a fault plane and other features that give an apparent fault separation but often erroneous indications of fault slip. Moreover, faulting may involve more than one episode, and the amount of displacement suggested by plate-kinematic considerations is not always evident in the rock record. Reconciling differences between predicted amounts of displacements, based on plate-tectonic reconstructions, and estimates of fault slip, based on field studies, has also revealed the importance of understanding *distributed shear*. Fault traces such as the San Andreas are single elements in a system of faults. Where these fault systems define plate boundaries the breadth of the boundary may be several hundred to several thousand kilometers wide. In this context, the upper crustal demarcation between the Pacific and North America plates embraces the entire region from the west escarpment of the Rocky Mountains to the base of the continental slope in the Pacific Ocean.

Thrust faults

The upper 10 km of the crust is fundamentally brittle; nonetheless, given a condition of excess hydrostatic pore pressure, sheets of strata may glide laterally near the Earth's surface, much like a hockey puck slides across the ice. The distances are sometimes impressive: 240 km has been estimated for the Canadian Rockies, 500 km for the Brooks Range of Alaska, 180 km for the Taiwan margin of China. The efficiency of thrust faulting may mask its very existence. Numerous examples, particularly of stratigraphic packages of mudstone and chert, were long believed to be intact depositional sequences. More recent paleontologic studies incorporating radiolarian biostratigraphy, however, have demonstrated repetitions of stratigraphic horizons. Thus, what once were thought to be conformable contacts are now interpreted to be thrust surfaces along which there has been many kilometers of horizontal slip; examples include Triassic and Jurassic chert sequences of the Mino terrane, Japan, and Paleozoic oceanic strata of the Golconda allochthon of Nevada.

In oceanic subduction zones, thousands of kilometers of ocean crust may slide beneath continental crust only to be lost into the mantle while the offscraped or underplated sedimentary cover gathers into stacks of thrust sheets. Where continental crust is shortened or where buoyant pieces of crustal material collide, sheets of crust become stacked upon one another forming architectural duplexes similar in configuration to the accreted sediment of subduction zones. In circumstances where the original stratigraphy was arranged in expansive layers it may be possible to reconstruct the amount of crustal shortening through a process called balancing. Imagine a pancake-like layering of stratigraphic units of an ancient continental platform or of a passive continental margin. Where these strata have become deformed into duplexes, they may be untelescoped by systematically stretching the shingles of strata back into their original inferred configurations. The resulting balanced cross section reveals the component of crustal shortening (Figure 5.21). Thin-skinned thrusting involves only the sedimentary strata. When chunks of basement ramp up into the duplex, so-called thick-skinned thrusting, a lesser amount of shortening is required to explain the observed crustal thickening. One of the enigmas in many crustal balancing exercises which deal with intracontinental shortening is the apparent loss of basement. If all of the observed crustal thickening is attributed only to thin-skinned duplexing, after reconstructing the sheets of strata into a presumed original configuration not enough basement may remain to support the extremities (the shingles placed end to end extend beyond the underlying basement). Does this mean that some buoyant continental crust has been subducted into the mantle, or does it require an element of thick-skinned duplexing, that is,

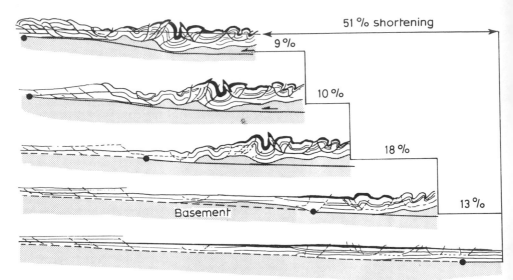

Figure 5.21 Cross section illustrating cumulative movement of a fold–and–thrust system during successive reconstructions in order to determine the amount of crustal shortening, a procedure called 'balancing'. The essence of this procedure is to account for the thickening of the crust. In this example crustal thickening is due to both folding and faulting; in the thickest part, basement is involved in the thrusting; such 'thick-skinned' faulting is more efficient in thickening the crust than 'thin-skinned' thrusting where only sedimentary strata are involved. An estimate of the amount of erosion is also included. If high-grade metamorphism had occurred, one would also need to account for more shortening owing to the diminished volume that accompanies mineral phase changes; anatectic plutonism (melting of sediment) also diminishes the crustal volume, whereas mantle–derived igneous rock will increase the volume, the latter requiring less and the former more crustal shortening than would be evident by simple balancing of the structures. Equating the percentage of shortening to distance depends on the scale of the cross section. Distances can range from millimeters to hundreds of kilometers. In this particular section, involving Paleozoic strata and Precambrian basement from Ireland, about 65 km of shortening has occurred. (Figure modified from Ford (1987))

was there less shortening, and some of the thickening results from the emplacement of slabs of basement? This is a difficult question to answer. One fruitful approach is to determine the extent of crustal flexuring in advance of the thrust sheets. If the crustal sag in the foredeep region is greater than the amount that could be caused simply by the weight of the thin–skinned duplexes, then some portion of the basement must clearly also have been involved in the stacking. Gravity modeling, seismic reflection, and seismic refraction may also reveal crustal thickening owing to the involvement of basement rock, but only in cases where the thrustbelts are younger than about 300 Ma, because, as was mentioned above, thick crust is gravitationally unstable, and the lateral flow of rock in the deeper ductile zones erases the roots of old mountain systems.

Balancing cross sections across a collage of terranes or within ancient subduction-related accretionary complexes is difficult, often impossible, because few or no reference horizons remain to balance. Paleomagnetic studies across Tibet indicate about 1500 km of crustal shortening, and some of this is probably a result of crustal duplexing. But to attempt a stratigraphic study in this region of accreted terranes, hoping to demonstrate a like amount of crustal compression, would probably be foolhardy.

Detachment faults

Detachment faults represent a special class of dispersion faulting; individual faults are generally low-angle surfaces with younger strata superposed over older strata, rather than the inverse of this as displayed with thrust faults. On a relatively local small scale, these can occur within a regime of strike-slip faulting, and on a larger regional scale, such faulting commonly follows an episode of compressional tectonics. In each of these occurrences the crust is thickened to the point of becoming unstable, and like an over-thickened pile of plywood or playing cards, the upper (younger) sheets begin to slide laterally exposing the lower (older) sheets. These gravity-driven detachments may slide 100 km, and crustal material from 10 km or deeper may become exposed in the core regions. Along the continental margin of the Iberian Peninsula ultramafic material from the mantle is exposed on the sea floor, presumably as a consequence of a crust-cutting detachment fault associated with an early phase of thermal uplift and rifting that developed into the North Atlantic Ocean. The metamorphic core complexes of the Great Basin of western North America are another expression of extensional detachment faulting and much of the Franciscan subduction complexes of California are also thought to have risen up from exposed anticlinal cores below a detachment fault (cf. Figure 2.2).

5.7 CONCLUSION

We know plates are moving. A variety of high-precision geodetic surveys as well as laser and astronomical telemetry experiments have measured small yearly displacements among several plates and across some plate boundaries. Modern seismicity indicates the location of plate boundaries, based on arrival times of seismic waves at three or more stations, and first-motion studies, based on the character of the seismic waves, suggest the sense of movement along those boundaries. But these real–time detections do not assist in reconstructing the past trajectories of plates or in estimating the displacement histories of terranes. For such results we must look for clues that are left in the geologic record, such things as paleomagnetic, paleontologic, and sedimentologic data that fingerprint a unique latitude on the spherical Earth, and geologic analyses that define truncated and offset features across faults. In all these studies we must understand not only the amount of displacement but also the timing of the tectonic events and the dynamic regime in which the displacement occurred. Terrane trajectories across the expanse of an ocean, or the translational slip along the margin of a continent may be rather straightforward; but these displacements are often masked by post-accretion dispersions involving large rotations, thrust, detachment, and strike-slip faulting. Unraveling these tectonic perplexities is the challenge of all terrane analyses.

6

Mountain building and the shaping of continents

6.1 OVERVIEW

Figure 6.1 outlines five areas selected to demonstrate mountain-building phenomena. Area 'A', the southwest Pacific, is a modern laboratory where a variety of ongoing plate-tectonic processes can be observed that will ultimately result in an orogenic system along what is now the eastern margin of Asia. Features of particular interest include the accretion of volcanic island arcs onto the margins of continents, as seen in the environs of Taiwan and Timor; the amalgamation of volcanic island arcs, as observed in the Philippine archipelago; A-subduction, as imaged seismically along the north margin of Borneo; and the slivering and dispersion of continental fragments, as viewed in the Banda Sea.

Area 'B', the Himalaya and the Tibet Plateau, illustrates head-on continent–continent collision where a large ocean basin previously existed in the intervening area. Area 'C', southern Europe and northern Africa, likewise exhibits continent–continent collision, but here only small basins, floored both by thinned continental and oceanic crust, existed between the continents, and a component of sinistral shear that contributes to the architectural style of the various mountain ranges is important (Pyrenees, Alps, and Apennines). Area 'D', the Cordillera of western North America, is the type area for illustrating the disposition of suspect terranes along a continental margin that has faced an open ocean since about 560 Ma. Mountain building in this region reflects complex relations among the processes of accretion, dispersion, and intracontinental shortening (A-subduction). Area 'E', the Andes of South America from Colombia to central Chile, comprises two markedly contrasting areas; the northern sector is illustrative of accretion tectonics primarily as a result of the obduction of oceanic crust, whereas the southern sector represents a collision margin with an anomalous absence of crustal accretion. In northern Chile, for example, the Andes constitute a continental-margin volcanic arc. The thick crust is primarily the result of intracrustal thrust tectonics, possibly associated with ridge push from the Atlantic side. To the west, within the subduction zone that defines the plate boundary with the Pacific plate, one discovers a history of tectonic subsidence and crustal erosion.

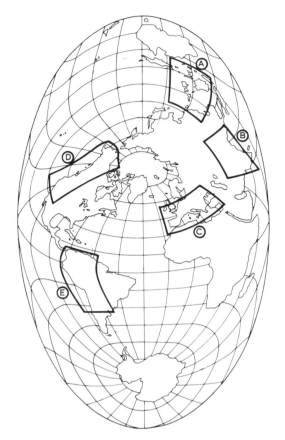

Figure 6.1 Index map locating five areas that will be discussed in this chapter. The primary interest is with the post-Paleozoic systems that evolved as a consequence of a new plate-tectonic regime that accompanied the breakup of Pangea: neotectonics of the southwest Pacific provide a living system of amalgamation, accretion, and dispersion; the Tibetan Plateau and Himalaya chain show the effects of microcontinental accretion associated with the closure of a major ocean basin in advance of a continent–continent collisional episode; the Alpine chains of Europe reflect styles of intracontinental subduction and the closure of small ocean basins; the Cordillera of the west margin of North America reflect the interactions between a continent and oceanic plates – crustal accretions followed by long-term coastwise dispersion; while the Andes evince two contrasting tectonic regimes: in the north oceanic material has been accreted on a large scale, in the south extensional tectonics and an apparent absence of accretion has accompanied the past 100 m.y. of subduction.

The salient features of each of the five regions are highlighted in order to underscore the principal styles of crustal thickening that result in mountain-building phenomena. To illustrate these styles, a host of cross-sectional renderings are presented. The reader should be cautioned that varying degrees of speculation are built into each section. A growing body of deep seismic data is now becoming available to constrain these notions further, but probably several decades will pass before geologists agree better on the origin of all these mountain systems.

6.2 TAIWAN TO TIMOR

The accretion of volcanic island arcs to the margins of continents is the most fundamental means to effect continental growth. Geologic and geophysical

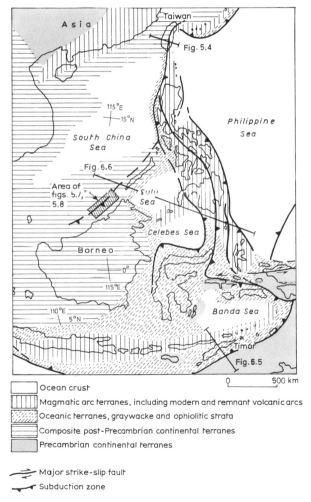

Figure 6.2 Generalized tectonostratigraphic terrane map for the southwest Pacific. Note the location of crustal profiles displayed in following figures.

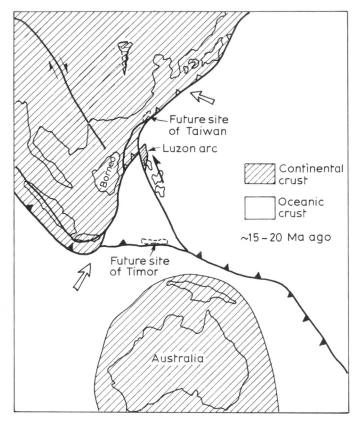

Figure 6.3 Generalized cartoon of the paleogeography of the southwest Pacific of about 15 Ma showing the distribution of continental and oceanic crust as well as the future sites of Taiwan and Timor. Taiwan resulted from the transpressional collision of a fragment of the Luzon arc with the continental margin of China; Timor, like the west part of Taiwan, is an uplifted part of the continental margin reflecting the collision with a volcanic arc (see Figures 6.4 and 6.5 for a cross-sectional rendering of the collision scenarios). (Modified from Jolivet, Huchon, and Rangin (1989))

investigations near Taiwan and Timor (Figure 6.2) have elucidated critical elements involving the processes of island-arc accretion. In terms of relative motions, these two accretions are conveniently viewed by holding the Eurasia plate fixed in space (Figure 6.3).

Accretion of island arcs

In the instance of Taiwan, a volcanic arc built on the oceanic Philippine plate has moved westward to the site of collision with the continental

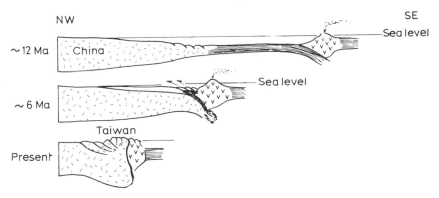

Figure 6.4 Simplified reconstruction displaying the crustal consequence of the collision of a fragment of the Luzon volcanic arc and the continental margin of China. The actual kinematics involved a major component of strike slip which is not obvious in this rendering. See Figure 6.3 for a plan view of the 15 Ma configuration. One must also realize that some of the oceanic crust that is being subducted eastward beneath the Luzon arc is young oceanic crust of the South China Sea. (Figure modified from Pelletier and Stephan (1986))

margin of the Eurasia plate (Figure 6.4). The Coastal Range along the east part of Taiwan is a remnant of the old arc, while the Central Range and regions to the west are deformed portions of the Eurasia continental margin. This collision has choked the east-directed subduction zone, and in the future, subduction will probably flip and become west directed. This configuration will connect the subduction zones of the Phillippine and Ryukyu Trenches, respectively south and north of Taiwan, effecting a continuous subduction zone along the west margin of the Philippine plate. Such a flip is also inferred for the Timor region, where the Wetar thrust may represent a nascent expression of the new subduction configuration (Figure 6.5). In this instance, a volcanic arc was built on the oceanic part of the Eurasia plate as a result of the north-directed subduction. The Australian continent has been moving northward since it broke away from Antarctica in the Late Cretaceous. Timor Island is composed of continental material that has been deformed owing to collision tectonics along the leading edge of the continental terrane as it descended into the old subduction zone. The inability of continental material to descend into the mantle effectively clogged the subduction zone resulting in the flip in polarity of subduction. With continued subduction Australia will ultimately collide with the Eurasian continent. The suture zone is likely to contain numerous terranes represented by the plethora of crustal fragments that now occupy the region of the Java, Banda, and Celebes Seas.

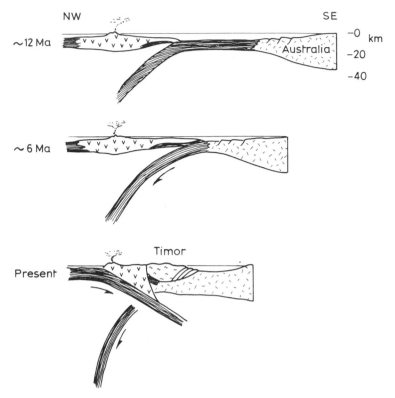

Figure 6.5 Cartoon showing the consequence of the collision of the northern trailing margin of Australia with the Banda volcanic arc. The island of Timor is a duplex structure composed of continental margin strata. (Figure modified from Price and Audley-Charles (1987))

Amalgamation and dispersion

The Philippine archipelago is a polyglot of terranes that represent continental blocks, remnant volcanic island arcs, backarc basins, ocean floors, and marine sediments (Figure 6.6). A variety of metamorphic assemblages indicates intraoceanic tectonic events. Superposed volcanic suites and overlapping sedimentary packages record earlier amalgamation events, and faults such as the Philippine fault continue to disperse the terranes into confusing geometric relationships that obfuscate previous geologic links. Little wonder that so much controversy surrounds the geologic history of this region. The dense jungle and political instabilities that have hampered field work are only minor nuisances compared with

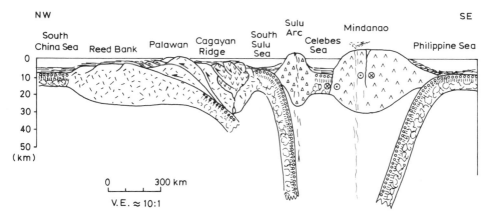

Figure 6.6 Crustal cross section traversing the Philippine archipelago from the South China Sea to the Philippine Sea. Features depicted include continental rifting along the west margin of Reed Bank, A-subduction within the microcontinental fragment of Palawan, the amalgamation of microcontinental fragments as evinced between Palawan Island and Cagayan Ridge, the growth of an island arc represented by the Sulu Arc, transpressive slip between the oceanic crust of the Celebes Sea and Mindanao Island, dispersion along the Philippine fault within the collage of arc terranes composing Mindanao Island, and west-directed subduction of the Philippine crust. This dynamic configuration is both ephemeral and unstable, characterizing the complexities inherent in a regime of active amalgamation and dispersion.

the intrinsic complexity of the region. The Philippine Archipelago provides an example that helps geologists to understand an intermediate stage in the evolution from intraoceanic amalgamations to the accretion of tectonic collages along continental margins. Doubtless most orogenic belts have had similar complex histories, which belies the simple paleogeographic renderings that often accompany efforts to reconstruct the tectonic evolution of a particular foldbelt.

A-subduction

The oceanic crust of the South China Sea is currently being subducted toward the east beneath the Philippines along the Manila Trench (Figure 6.2). This trench expresses active subduction from just south of Taiwan to Mindoro Island. South of Mindoro, however, no simple solution is evident regarding the continuation of the plate boundary. In the Pliocene, it may have continued along the west margin of Palawan Island, but sediment now drapes across the fossil trench, demonstrating a present-day absence of subduction. An alternative explanation involves a complicated transform system that links subduction beneath the Sulu arc and the Manila Trench, but how this subduction is linked to the other microplate boundaries of the Indonesian archipelago is also not understood. Nonetheless, along the northwest margin of Borneo seismic reflection profiles reveal thrusting that must be integrated in this plate-tectonic scenario (Figure 6.7). The thrusting seems to be a southward continuation of the now–dormant subduction zone west of Palawan Island. The floor of the southern part of the South China Sea, in the region of Reed Bank and Dangerous Ground, is composed of thinned continental crust, presumably formed during the early to middle Tertiary extension of the east margin of Asia that resulted in the formation of this sea. Borneo itself is essentially continental in character, a part of Sundaland, with a Paleozoic basement and thickened as a result of Tertiary thrusting. Thus, the crustal shortening shown in Figure 6.7 is wholly within a continental domain.

The age relations of the thrusts along the north margin of Borneo are magnificently recorded by the stratigraphic character of *piggyback basins* (Figure 6.8). These are the small basins on the surface of a thrust duplex that formed behind each successive ramp anticline. In accretionary wedges associated with B–subduction (Wadati–Benioff) zones, such basins are commonly referred to as *trench-slope* basins. In the case of the Borneo examples, the more southerly basins clearly have a thicker basin fill, and the lower stratigraphic horizons of this fill are tilted toward the south. This indicates that the basins and the responsible thrust faults are progressively older toward the south, that is, this foreland fold-and-thrust belt is advancing toward the north. By drilling into the sedimentary packages of the piggyback basins, one could establish the precise chronology of thrusting.

Continental slivers

Numerous islands and submarine ridges in the Banda Sea region have a continental foundation (Figure 6.9). The history of each is still not well known, but a reasonable interpretation is that they represent fragments of the Irian Jaya portion of continental Australia. Previous episodes involving a component of transcurrent motion between the Australia and Pacific plates have torn fragments of continental crust away from this promontory of the Australian continent. But a fair question to ask is why did the transcurrent faults bite off a portion of continental crust rather than slice along the continent–ocean plate boundary? The answer has to do with the fundamental strength of rocks and is nothing more than the fact that quartz, an abundant constituent of continental crust, is weaker than olivine,

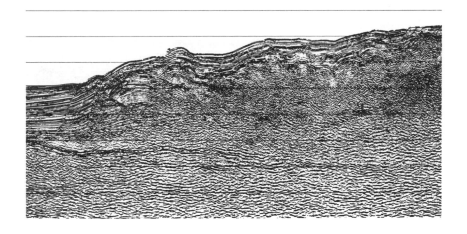

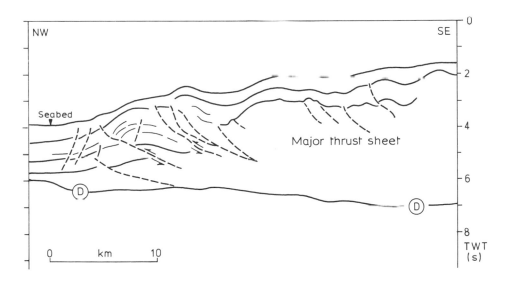

Figure 6.7 Multichannel-seismic (MCS) line across the Palawan Trench northwest of the island of Borneo. This represents a fold-and-thrust belt in an A-subduction regime. Horizon D is the top of the attenuated continental crust that has rifted from China. The material involved in the duplex structure is Miocene and younger sediment deposited in the young South China Sea. (Seismic line and interpretation compliments of Karl Hinz and the Bundesanstalt for Geowissenschaften und Rohstoffe, Hanover, Germany)

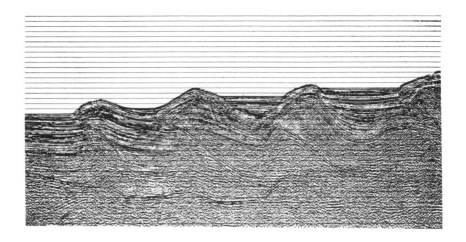

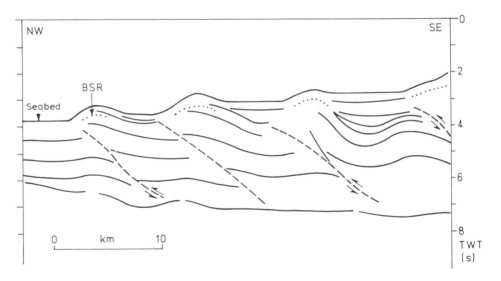

Figure 6.8 MCS line across the Palawan Trench northwest of the island of Borneo. Deformation is within Miocene and younger strata lying above the basal decollement believed to be the top of attenuated continental crust that has rifted from the continental margin of China. The northwestward propagation of thrust faulting is evinced by the character of the overlying 'piggyback' basins. Stratal fill is thickest in the southeast, and lower horizons in the southeasternmost basin are themselves folded as a consequence of the growing ramp anticline. The precise age of faulting could be determined by a stratigraphic analysis of boreholes drilled in these piggyback basins. (Seismic line and interpretation compliments of Karl Hinz and the Bundesanstalt for Geowissenschaften und Rohstoffe, Hanover, Germany)

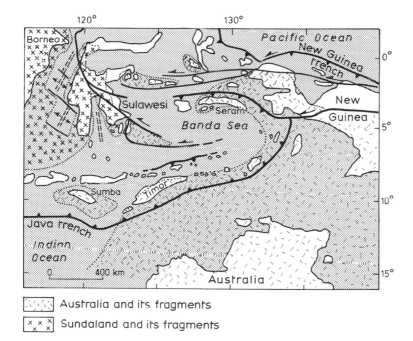

Figure 6.9 Simplified tectonic and crustal map of the Banda Sea region showing the distribution of dispersed microcontinental fragments that lie sandwiched between the continental masses of Australia and Sundaland. (Figure modified from Hamilton (1979))

the principal mineral composing oceanic crust. This strength differential is compounded at higher temperature, which explains why thick continental crust is weaker than thin continental crust. Thus, continental lithosphere is weaker than oceanic lithosphere, and the weakest zone in a stress regime yields preferentially to the stress (Figure 6.10). Rifting close to an ocean–continent boundary follows a continental pathway, and this characteristic helps account for the continental slivers in the Banda Sea region and the abundance of such terranes in the tectonic collages that mark older suture zones.

6.3 HIMALAYA AND TIBET PLATEAU

The high plateau of Tibet and the Himalayan foldbelt (Figure 6.11) are the manifestation of the early Tertiary collision between the Archean craton

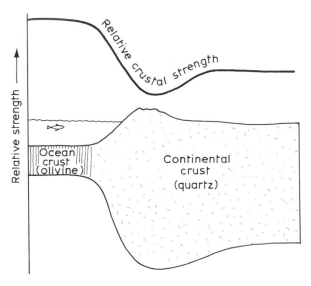

Figure 6.10 Crustal rock is strengthened by overburden pressure but weakened with elevated temperature. Quartz, which has a lower crystallization temperature than olivine, is fundamentally weaker than olivine, and therefore continental crust is weaker than oceanic crust. Furthermore, thick continental crust is weaker than thin continental crust because of the higher temperature in the lower reaches of the thicker crust. This figure depicts a qualitative rendering of the relative strength of the crust across an ocean–continent transition and explains why most rifts follow a continental pathway. Shear across an ocean–continent boundary will produce microcontinental fragments. See Vink *et al.* (1984) for a more rigorous treatment of the physical properties and strengths of continental and oceanic crust.

of India and the south margin of Eurasia that is itself composed of previously accreted volcanic arcs and smaller continental fragments. The general style of the initial phase of this collision is different from that observed at Timor and Taiwan only in terms of scale (Figure 6.12). The size of the collision, however, effected a global consequence that is not evident in smaller collisions such as at Taiwan or in the collisions along the south margin of Eurasia that presaged the arrival of India. The accretion of India was a crash felt around the world. In the Pacific, a major plate reorganization at 43 Ma, as recorded by the abrupt bend in the Hawaiian–Emperor seamount chain, seems to be linked to this collision. In the North Atlantic, at about the same time, the midocean spreading jumped from a ridge in the Labrador Sea to the Reykjanes Ridge east of Greenland. Many of the major volcanic arcs in the Pacific began to form at

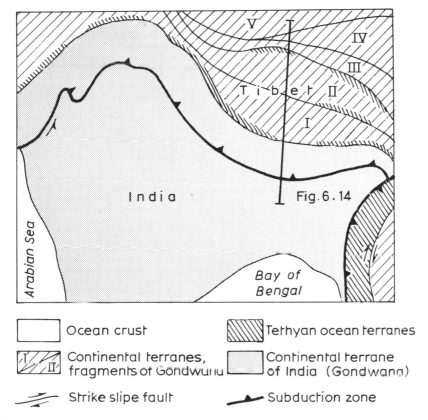

Figure 6.11 Generalized terrane map of the Tibetan Plateau and Himalayan orogen.

about 50 Ma (Tonga, Mariana, and Aleutian), although the precise nature of cause and effect between the initiation of new subduction zones and the collisional tectonics in India and Eurasia remains uncertain. Somewhat better understood is the nature of deformation within Eurasia as a consequence of this collision.

Prior to India's collision with the south margin of Eurasia the India Plate was moving northward relative to the Eurasia plate at about 7 cm/yr. The remarkable aspect is that after the collision India did not stop but only slowed down, to approximately 5–6 cm/yr. Paleomagnetic inclination data have confirmed the approximate 1500 km of northward penetration into the Eurasian continent. The absorption of India into Eurasia is accommodated by three kinematic expressions: intracontinental shortening by folding and thrusting, expulsion of continental material toward the

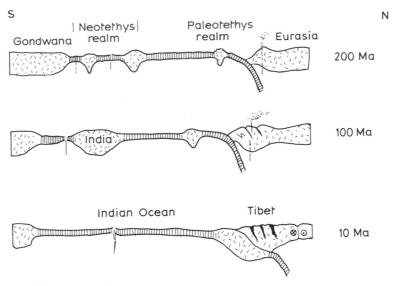

Figure 6.12 Cartoon depicting the accretion tectonics along the south margin of Eurasia effecting the thickened crust beneath the Tibetan Plateau. See Figures 4.7, 4.16, and 5.13 for complementary aspects of these collision-accretion events. (Figure modified from Tapponnier *et al.* (1981))

west and east of the advancing prong of India, and A-subduction effecting a doubling of the continental crust.

From Figure 6.11, the structural grain of Eurasia clearly wraps around India the way a rug might yield to a foot forced into an edge, forming sweeping bow-wave-like festoons. This horizontal shortening, and the consequential vertical thickening, surely contribute to the great thickness of the Tibet Plateau; but at the present time, no agreement exists among the area experts on just how much of the 1500–2000 km of northward movement of India is accommodated in this manner.

Because the south margin of Eurasia is composed of numerous terranes recording accretion events during the closure of the various Tethyan oceans, numerous suture zones constitute fundamental zones of weakness. Thermal softening associated with the north-directed subduction further weakens the Eurasian margin; thus, as India impinged against the south margin of Eurasia, many of these old suture zones became reactivated as strike-slip faults. Blocks in the Afghanistan region are expelled to the west, while most of China has moved eastward (refer back to Figure 4.16). Across the Tibetan Plateau, Quaternary N–S oriented pull-apart basins are a manifestation of modern E–W extension. First-motion seismic analyses

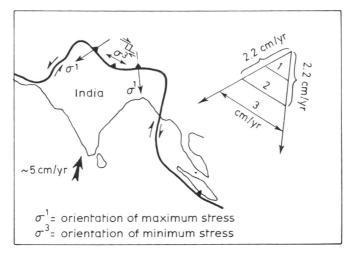

Figure 6.13 Recall Figure 5.3 where earthquake seismicity defines the orientation of crustal movements. The azimuthal orientation for thrust faults along the suture of India and the Tibetan Plateau converges toward the north. The Tibetan Plateau also evinces crustal rifts that are extending normal to the principal direction of compression. A recent rate of slip of approximately 2 cm/yr has been estimated for NW−SE rifting by dating offset soil horizons and geomorphic features. Combining the rifting vector with the two thrust vectors makes it possible to calculate the recent rate of northward movement of India relative to Tibet. The approximate 2 cm/yr solution represents 40% of the relative motion of the India plate relative to the Eurasia plate. Thus, India is being subducted beneath Tibet and, additionally, the Tibetan Plateau is moving northward relative to the cratonal portion of the Eurasian plate. This latter motion is accommodated by the west- and east-directed extrusion of crustal fragments throughout the south margin of Eurasia, effectively making room for the forward advance of India and Tibet. See Figure 4.16 for a rendering of the extrusion kinematics along the east part of China. (Figure modified from Armijo *et al.* (1986))

from earthquakes as well as the general map patterns of major thrust faults indicate a convergence which on a vector diagram seems to emanate from a single point (Figure 6.13). Combining the estimated 2 ± 0.5 cm/yr of Quaternary slip for the E−W pull-apart basins with the orientation of the thrust faults allows one to calculate an approximate 2 to 2.5 cm/yr of closure between India and the Tibetan plateau. This convergence is a manifestion of only very recent kinematics; however, other estimates from the south-verging thrusts north of the Ganges Plain, which record the longer-term closure of India and Tibet Plateau, are consistent with this rate.

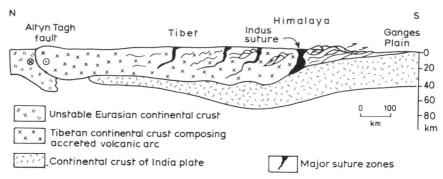

Figure 6.14 North–south cross section depicting India underthrusting the south margin of Asia.

But the India plate is converging on the Eurasia plate at a rate of between 5 and 6 cm/yr. The remaining 3 to 4 cm/yr of motion between India and the craton of Eurasia must therefore be absorbed by folding and thrusting within the Tibet plateau and the indenting of an amalgamated India–Tibet ensemble into the south margin of Eurasia, causing a lateral fleeing of crustal subplates to accommodate this northward motion.

Thus, even though India has continued to move northward by as much as 1500 km since the initial contact with the south margin of Eurasia, it is kinematically not required to subduct continental crust into the mantle. Figure 6.14 is a speculative cross section where 1500 km of the north movement of India is reflected in 700 km of A-subduction, resulting in a 'doubling' of the thickness of continental crust, 300 km of intra–plateau thickening (thrust faulting and folding), and the remaining 500 km registered by the crustal-expulsion phenomena.

The forces involved in the dynamic setting of the Indian collision remain enigmatic. The crust below the Ganges Plain is Early Archean in age, whereas the lithosphere beneath the south margin of Eurasia is principally Mesozoic in age. This pronounced difference in age may result in enough contrast in density between the two continental lithospheres to permit a component of slab-pull. The ocean crust south of India has also buckled into several E–W oriented anticlines indicating ridge-push forces as well. Possibly, as the effectiveness of the slab-pull forces diminishes, ridge push forces cause the trailing oceanic plate to fail structurally. In the future, a new subduction zone may evolve, trapping India and a major part of the Bengal Sea, in a manner similar to the trapping of Shirshov Ridge and oceanic crust of the Bering Sea north of the Aleutian arc.

6.4 AFRICA–EUROPE COLLISION

During the late Paleozoic, Africa (the northern part of Gondwanaland) collided with Eurasia and Laurentia in the final consolidation of the supercontinent of Pangea. This collision is known in Europe as the Hercynian orogeny and in North America as the Allegheny orogeny. Numerous Hercynian sutures traverse Europe from southern England to southern Spain, suggesting a protracted period of collision tectonics similar in style to the younger episodes of accretion that preceded the arrival of India at the south margin of Eurasia. Following the period of crustal thickening of the European sector, the crustal compression was relaxed in the Permian, and numerous extensional basins were superposed across the Hercynian sutures. Late Jurassic and Cretaceous left-slip transform faulting along the area of the Hercynian sutures was linked to the initial opening phases of the North Atlantic. This resulted in the formation of pull-apart basins that extended the Neo-Tethyan ocean to the west, allowing a seaway to connect the Atlantic with the Tethys. The post-Hercynian paleogeographic evolution is not well known, because in the Tertiary, Africa and Europe again began to collide in what is known as the Alpine orogeny. In unraveling the Alpine tectonics of the Pyrenees, Alps, and Apennines one must consider these earlier tectonic events, as the stage that they set controlled many of the features that characterize the Alpine orogeny (Figure 6.15). In the simplest rendering, the Pyrenees represent crustal thickening within the Europe plate; in the Alps this thickening involves the joining of both the Europe and Africa plates; whereas in the Apennines, the thickening is wholly within the Africa plate. How does all this happen?

The Pyrenees

An overall cross-sectional fan shape to the Pyrenees reflects the collision of two continental crustal blocks effecting thrust systems that verge toward the north and toward the south. The five principal tectonic features from south to north that characterize this range are:

(1) the Ebro basin, an Eocene to Miocene foreland for the south-verging thrusts that are exposed along the southern flank of the Pyrenees;

(2) the Axial zone that forms the highest part of the range and comprises reworked Hercynian basement complexes;

(3) the near-vertical, E–W trending North Pyrenean fault zone (NPfz) that seems to be a sinistral system associated with mid-Cretaceous opening of the Bay of Biscay, i.e. left slip and counterclockwise rotation of the Iberian crustal block away from

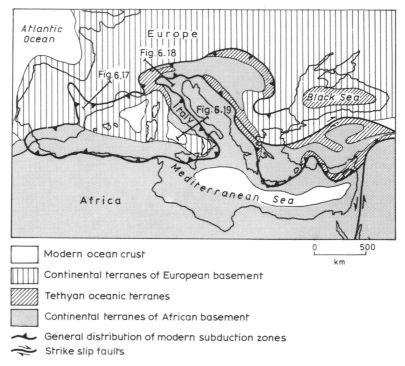

Figure 6.15 Generalized tectonostratigraphic terrane map for southern Europe. (Modified from Aubouin (1984))

the European crustal block that crops out along the coast of Brittany;

(4) a region of north-verging thrust faults that cuts across an area extending to the north from the NPfz with Cretaceous flysch that grades into coarse-grained nearshore deposits that lap onto Hercynian basement highs; these Mesozoic sedimentary sequences are interpreted to be pull-apart basin deposits associated with left slip along the NPfz (Figure 6.16), and similar Permian sequences are inferred from seismic-reflection data, but these deposits do not crop out in the range; and

(5) the Aquitaine basin, a latest Cretaceous to Oligocene foreland for the north-verging thrust system cropping out along the north margin of the Pyrenees.

Other geologic relations are important in this mountain chain, but the features enumerated above, in conjunction with a newly acquired deep

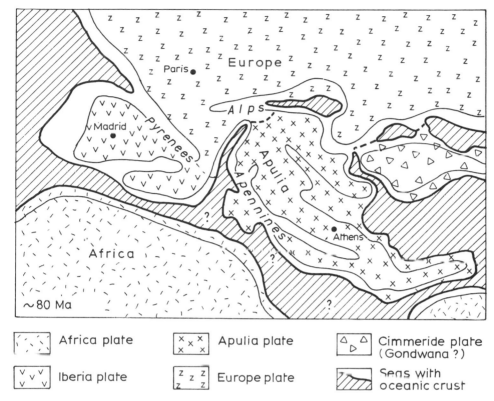

Figure 6.16 Speculative paleogeographic map of Europe for the period around 80 Ma depicting the general distribution of marginal seas, areas of oceanic crust, and distinct continental crustal fragments. The future sites of the Pyrenees, Alps, and Apennines are also indicated; these mountain systems reflect varying manifestations of the collision between Africa and Europe. (Figure modified from Dercourt *et al.* (1986))

seismic-reflection profile (Figure 6.17), provide the critical constraints necessary to interpret the Alpine orogeny for this region. Following the episode of left slip along the North Pyrenean fault zone, Africa and Europe began to close upon one another. The thinner European crust was the first to yield to compressional tectonics. The south margin of this Hercynian crustal fragment thickened by folding and north-verging thrusting as evidenced by the stratigraphy of the Aquitaine basin. By Middle Eocene time, the south-plunging A-subduction zone gave way to north-directed subduction of the Iberia plate. This chronology is reflected in the stratigraphic relations of the Ebro basin. Balanced cross sections, which

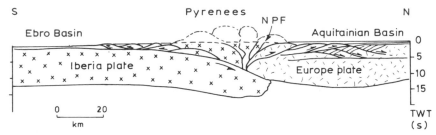

Figure 6.17 Schematic cross section across the Pyrenees depicting 100 km of crustal closure between the Europe and Iberia plates. These two plates were initially accreted in the late Paleozoic during the Hercynian orogeny. Mid-Cretaceous left-slip transtensional rifting reactivated this suture. North-directed normal faults on the Europe plate are an expression of this tensional regime. By latest Cretaceous time compressional tectonics resulted in north-verging thrust faults and the development of the Aquitainian foreland basin. With continued compression, the direction of thrusting reversed, creating the Ebro foreland basin by Eocene time. (Figure modified from Choukroune, Lopez-Munoz, and Quale (1983))

restore the stratigraphy of the two foreland basins and account for the crustal thickening in the axial part of the range, indicate a total shortening of approximately 125 km. Evidence suggests that no continental crust has been subducted into the mantle. The Permian and middle Cretaceous crustal thinning events reduced the thickness of the Europe plate to an extent that, with the aforementioned shortening, crustal thickening that can be accommodated in the volume made up by the root zone and that part of the range that has been eroded off since the initial phases of thrusting.

The Alps

In most respects, the European Alps are the birthplace of tectonics. It was here that during the late nineteenth and early twentieth centuries, concepts such as the thrust fault, the allochthonous nappe, and A-subduction were first conceived in order to explain the confounding contortions of strata and enigmatic crustal relations evinced in this spectacular landscape. However, without trivializing the hundreds of geologist-years of effort devoted to working out the geometric patterns of the disrupted strata, the Alps can be characterized by four basic units whose relations reflect a simple duplex structure resulting from south-directed subduction and the closure of small oceanic basins between two continental masses.

Figure 6.16 portrays the fundamental features that presaged the Alpine orogeny. After the closure of Gondwanaland and Laurasia, the North Atlantic began to open in Middle Jurassic time. Ocean-ridge spreading in the Tethys and North Atlantic were connected via a complex and as yet incompletely understood system of left-slip transform faults. Within this fault system an unknown number of pull-apart basins developed, some evolving into marginal seas floored by ocean crust. By the middle of the Cretaceous, transpressive tectonics initiated a protected phase of basin closure, and by the latest Cretaceous, closure between Europe and Africa was largely orthogonal to the conjugate margins of these two continental masses.

The following are the four principal elements of the Alps and adjoining regions that record their history (Figure 6.18), discussed from north to south, or, tectonically, approximately from youngest to oldest:

(1) the Jura Mountains, a fold-and-thrust belt of late Tertiary age that lies above a major blind thrust, specifically the youngest thrust in the Alpine system that has yet to ramp up to the surface; the disrupted strata are part of the European platform and older post-Hercynian basinal deposits;

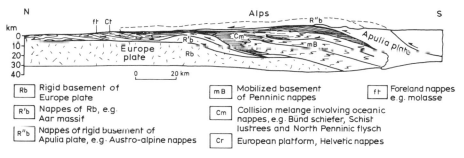

Figure 6.18 Stylized rendering of a N–S cross section through the Alps. The principal components are two continental blocks and material from an intervening marginal sea with Jurassic oceanic crust. The northern, European continental block is the lower plate of this duplex structure. Deep crustal rocks reflecting an older Hercynian orogenic event have been transported upward and to the north; these are the so-called remobilized blocks composing the crystalline Penninic nappes (or terranes). Structurally admixed with the crystalline Penninic nappes are terranes composed of Jurassic oceanic crust and younger basin-filling flysch. To the north, higher-level basement-involved thrusting accounts for nappes such as the Aar massif. The platformal strata that once lay on this crystalline basement compose the Helvetic nappes. A blind thrust to the north has wrinkled the overlying strata as reflected in the folds of the Jura Mountains. The southern continental terrane of Apulia has compressed and overridden most of the aforementioned thrust nappes. (Figure modified from Hsü (1981) and Funk *et al.* (1987))

(2) the middle Tertiary Helvetic zone composed of rigid slices of Hercynian basement, thrust nappes involving Mesozoic strata from the European platform, and foreland deposits represented by the molasse of the Swiss Plateau;

(3) the Late Cretaceous to middle Tertiary Penninic zone comprising remobilized deep-crustal Hercynian basement nappes that are engulfed in a sea of graywacke and ophiolitic terranes representing middle Mesozoic oceanic basins; and

(4) the Austro-alpine nappes that are basement fragments of the African continent. Klippes of the Austro-alpine nappes (see for example Figure 4.4) are found as far north as the Helvetic zone indicating that the overriding crustal wedges of the African basement are at least in part a late-phase event. The middle to late Tertiary movement along the basal thrust is an out-of-sequence phenomenon stimulated by continued compression between Africa and Europe long after the closure of all intervening oceanic basins.

The total amount of crustal shortening registered by the Alpine orogeny is difficult to estimate for two reasons: firstly, because no well stratified foreland deposits are available, such as in the flysch of the Ebro and Aquitaine basins that flank the Pyrenees, to record the entire episode of thrusting and to themselves be shortened by the propagation of the duplex system of thrusting; and, secondly, because the original breadth of the small oceanic basins that occupied the region between Europe and Africa prior to the Alpine orogenic closing event is not known. Paleomagnetic data, with a minimum resolution of $3°$ (roughly 330 km), are unable to distinguish differential movement between the two plates; therefore, the maximum closure between these plates must be less than about 300 km. The Austro-alpine nappes evince at least 125 km of northward thrusting and published estimates of shortening within the Helvetic and Penninic nappes are between 100 and 200 km; hence a reasonable amount of total crustal shortening for the entire Alpine system is probably in the range of 200 to 300 km.

The Apennines

During the Jurassic phase of left-slip transform faulting that sliced through the modern Mediterranean region, a large chunk of Africa broke off the main continental mass and became the microcontinent of Apulia (Figure 6.16). The northern flank of this microcontinent is what makes up the rocks of the Austro-alpine nappes. During the process of unhinging of Apulia from Africa, numerous areas of the microcontinent thinned, forming intracontinental basins and zones of weakness that would later yield bewildering kinematic expressions, caught in the vise made up of the African and European plates. One pair of manifestations of this complex kinematic history is counterclockwise rotation resulting in Miocene extension in the Ligurian Sea north of Corsica, and Plio-Pleistocene opening of the Tyrrhenian Sea between Sicily and Sardinia.

Contemporaneous with the rotation of all of the Italian Peninsula was an accordion-like stacking of crustal sheets along east-directed thrust faults. This crustal shortening has resulted in the formation of the Apennines. In southern Italy, in the states of Abruzzo and Molise, four basic units can be distinguished that provide the critical relationships necessary to understand the mountain-forming processes of the Apennines. From west to east these are (Figure 6.19):

(1) a sequence of high-level nappes that comprise lithofacies believed to represent a western carbonate platform;
(2) middle-level nappes that are made up of deep intracontinental basinal deposits;
(3) the structurally lowest nappes that consist of carbonate platform sequences that are separated in a paleogeographic sense from (1) in the west by the deep basinal deposits of (2); and
(4) the modern foreland deposits for this thrust system which is still active.

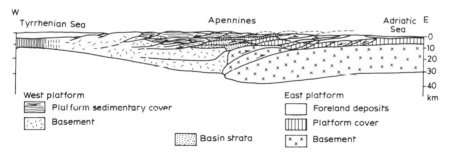

Figure 6.19 Schematic E–W cross section across the Apennines showing the oceanic crust of the Plio–Pleistocene Tyrrhenian Sea on the west and the effects of A-subduction within the Apulia plate to effect the Apennine fold-and-thrust chain. The principal thrust nappes consist of two Cretaceous platformal sequences and an intervening deep-basinal facies; see the prethrusting paleogeography of Figure 6.16. Not seen in outcrop but inferred in this rendering are deep-seated thrusts that involve crystalline basement. The effect of this is to reduce the amount of thrusting required to account for the thickened crust. Coeval with thrusting are several events of crustal extension, presumably a consequence of overthickening of the crust. (Figure modified from Roure *et al.* (1989))

A puzzling aspect of the Apennines is why thrusting is directed eastward when the principal orientation of compression is N–S. One explanation attributes the thrusting to the counterclockwise rotation of the Italian Peninsula and likens it to the folding of a pleated paper fan. An alternative scheme envisages a N–S squeezing of the crust, which deforms by sliding to the east in the manner that a thin sapling bows outward when its ends are forced toward the center. In both of these scenarios, the eastward-migrating thrust regime is coupled to the opening of the Ligurian and Tyrrhenian Seas along the west flank of the Apennines. Still another possible mechanism relates this thrusting to the expulsion of the Anatolian and Aegean region to the west as a consequence of the compression due to the counterclockwise rotation of the Arabian Peninsula into the southern margin of Eurasia. And a final kinematic scheme envisages that the descending Africa plate is rotating counterclockwise about a pole of rotation located near the 'heel' of Italy's boot. In each of the last two scenarios the young oceanic crust west of Italy is formed by backarc-basin spreading. A consensus solution must await further study. For now we must be content simply with an understanding of the kinematics that resulted in the formation of the Apennines.

The Apennine thrust system is a classic example of A-subduction. In the region of the cross section, all the thrusting has been intracontinental (Figure 6.19). The floor of the Adriatic Sea is inferred to be thinned continental crust, so as the system continues to propagate to the east the character of thrusting should persist. The aggregate amount of crustal shortening based on balanced cross sections, such as displayed in Figure 6.19, is approximately 100 km. In the south of Italy where the Apennine chain swings to the west, crossing Sicily and extending toward Tunisia, the subducted slab is believed to be oceanic. Modern volcanoes such as Mt Etna and Mt Vesuvius are seemingly linked to this regime, which is one of B-subduction.

6.5 CORDILLERA OF NORTH AMERICA

The mountainous aspect of the Cordillera of North America is primarily a post-Paleozoic phenomenon. Beginning in the latest Proterozoic to earliest Paleozoic at 600 to 550 Ma, continental rifting along the west margin of the North American continent initiated a protracted passive-margin period. Other than a poorly understood Late Devonian to Early Mississippian (360–340 Ma) tectonic disturbance that is recorded in the stratigraphy from the Arctic Islands to Nevada, the tranquil trailing-margin setting persisted until the Triassic, coincident with the transition from the assembly to the breaking up of Pangea. This latter event marks a time of global reorganization of plate kinematics, and the effects in the Cordillera were profound.

Three cross sections traversing the west margin of North America have been selected to illustrate different aspects of the Cordilleran orogenic belt (Figure 6.20). The Alaskan example involves numerous tectonic styles. The entire State is composed of allochthonous crustal pieces, from far-traveled oceanic plateaus to nearby continental slivers, most of which are encased in a graywacke matrix. Southern British Columbia consists of volcanic-arc sequences accreted in the Jurassic and Cretaceous, but the fold-and-thrust tectonics that is so well displayed across this margin is primarily an A-subduction phenomenon associated with ridge-push forces from the Atlantic. In the area of the United States–Mexican border, dispersion tectonics has been particularly important. Nearly all of the Paleozoic 'miogeoclinal' sequences have been stripped from this area. In the Middle Jurassic, left-slip transform faulting that connected oceanic spreading in the Gulf of Mexico to subduction zones in the Pacific traversed this region causing major dislocation of the Precambrian basement terranes. Slivers of continental-margin volcanic arcs have been transient features along this region, reflecting a combination of transpressional and transtensional tectonics along with arc-generating subduction.

Alaska

The topography as much as any other feature indicates the youthful nature of Alaskan geology. In the south, bordering the Gulf of Alaska and extending inland for 300 km, is the precipitous Alaska Range, where peaks are 5000 to 7000 m high. The stratigraphic record in the Gulf of Alaska indicates that this greatly elevated topography is probably younger than 5 Ma. To the north of this rugged terrain, in the middle of the state, the university city of Fairbanks is only 120 m above sea level. Northeast of here, in the triangular region between the Porcupine and Yukon Rivers, elevations average about 100 m, and the only topographic relief in this 60000 km^2 region is that of incised riverbanks – by comparison, the state of Kansas seems mountainous. Across the entire northern edge of Alaska is the Brooks Range, an approximately 100 m.y. old mountain system that has retained a youthful countenance. Elevations increase gradually toward the north to a crest that averages about 2500 m. The north face is steeper as the range descends to the 100–150 km wide tundra swamps of the Colville Plain.

Accretion in Alaska, like the topography of her mountains, is oldest in the north and becomes progressively younger to the south. All of Alaska, as we know it today, has formed since the middle of the Cretaceous (refer to

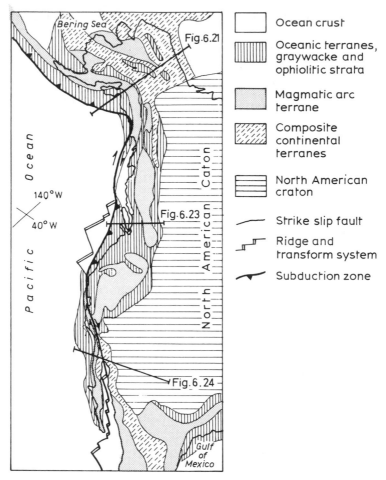

Figure 6.20 Generalized terrane map for a major portion of the Cordillera of North America.

Figure 4.13). A brief but hopefully not oversimplified account of the tectonic history of Alaska includes the salient elements to follow (see Figure 6.21 for a cross-sectional rendering of these features).

Brooks Range

The Brooks Range is a fold-and-thrust belt involving at least 500 km of crustal shortening. Thrusting began in the Late Jurassic and essentially terminated by the middle of the Cretaceous even though local compressional features are as young as latest Pleistocene. Four compo-

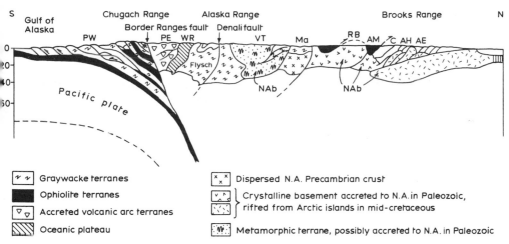

S Gulf of Alaska

Chugach Range   Alaska Range   Brooks Range   N
Border Ranges fault Denali fault

PW   PE WR   VT Ma   RB AM C AH AE

Flysch

Pacific Plate

NAb   NAb

	Graywacke terranes
	Ophiolite terranes
	Accreted volcanic arc terranes
	Oceanic plateau

	Dispersed N.A. Precambrian crust
	Crystalline basement accreted to N.A. in Paleozoic, rifted from Arctic islands in mid-cretaceous
	Metamorphic terrane, possibly accreted to N.A. in Paleozoic

Figure 6.21 Schematic N–S cross section of Alaska from the diffuse subduction in the Gulf of Alaska to the trailing margin along the Beaufort Sea of the Canada Basin. Compositionally, most of the terranes represent ocean crust and overlying deep-sea clastic lithofacies, accreted volcanic arcs, or crystalline fragments transported from the edge of the North American continent. The assembly of this diverse group of terranes has happened since the middle of the Cretaceous. The emplacement of the Brooks Range (see Figure 6.22) created a backstop against which the terranes to the south have accreted. The kinematics of accretion have involved an interplay between northwest trending transpressional tectonics and more orthogonally oriented B-subduction. (Figure modified from Coney and Jones (1985))

site terranes make up the stratigraphic framework of this range. The structurally lowest Endicott terrane includes an amalgamated assemblage of pre-Devonian metamorphosed volcanic-arc terranes that is the basement for a Late Devonian to Late Jurassic continental-margin sequence. Structurally above the Endicott are numerous thrust slices of the Hammond terrane consisting of thick units of lower and middle Paleozoic carbonates, schistose clastic and volcanic rocks, and Devonian plutons. A few radiometric dates suggest a Precambrian basement. This confusing assemblage of continental-margin strata is overlain by a Late Devonian clastic sequence. The next higher assemblage of nappes includes polymetamorphosed clastic strata of the Coldfoot terrane. The age and depositional setting of these strata are unknown. Capping the entire sequence of thrust nappes are Mississippian to Jurassic ophiolitic and clastic strata of the Angayucham terrane.

The age for thrusting is bracketed by the emplacement of Jurassic strata in the highest nappes and foreland deposition involving latest Jurassic to middle Cretaceous strata in the Colville basin. But much doubt and many controversies remain regarding the pre-thrusting paleogeography. Three models have been invoked to explain the present orientation of the Brooks Range (Figure 6.22):

(1) post-thrusting large-scale right slip along the British Columbia margin of North America, dispersing the Brooks Range sliver to the northwest;

(2) left slip along a southwest shear zone, penecontemporaneous with thrusting, sliding the Brooks Range away from the Arctic Islands; and

(3) post-thrusting counterclockwise rotation of the Brooks Range away from the Arctic Islands.

Lying directly to the south of the stacked sequence of nappes composing the Brooks Range is a Lower Cretaceous volcanic-arc terrane. An appealing kinematic model employs analogies from Taiwan and Timor where volcanic island arcs have collided with a continental margin effecting a series of continentward-verging thrusts. The vast lateral extent of thrusting in the Brooks Range, however, seems too great for such a simple collision. An alternative is to draw upon A–subduction processes forced by the Early Cretaceous initiation of spreading in the North Atlantic. An entire tome could be devoted to the pros and cons of all the variously proposed kinematic models. The reference noted in the caption to Figure 6.22 provides a good starting place for any interested reader wishing to pursue this topic. Further clarification, or even enumeration, is beyond the scope of this book.

Alaska Range

By whatever means the Brooks Range was emplaced into northern Alaska, it was accomplished by the middle of the Cretaceous, because at that time the extensional Canada basin north of it had formed. The effect of the Brooks Range was to create a backstop against which terranes that were moving northward within the Pacific realm could accrete. The large right-slip faults of the Tintina and the Denali–Fairweather fault systems reflect dispersion tectonics involving the transport of terranes along the western margin of North America. Some of these terranes are slivers of continental North America, others are exotic terranes that accreted to North America in more southerly latitudes. The west margin of North America probably resembled the modern borderland region of the southwest Pacific, north of Australia and between Irian Jaya and Borneo. And just as the basins there are receptacles for thick accumulations of sediment, the abundance of graywacke amongst the terranes of southern Alaska evinces a paleogeo-

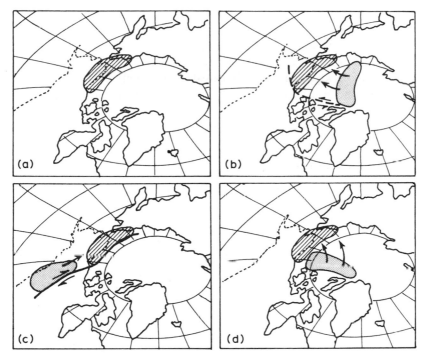

Figure 6.22 Cartoon depicting three kinematic scenarios that have been published to explain the emplacement of the Brooks Range and correlative units in northeastern Russia: (a) present-day configuration; (b) a left-slip model where the Brooks Range rifts away from the Lomonosov Ridge of the Arctic Ocean; (c) a right-slip model where the Brooks Range strata are slid northward from a Cordilleran source; and (d) a counterclockwise rotation model where the Brooks Range strata are moved away from the Arctic Islands. Preliminary paleomagnetic data seem to support scenario (d). See Howell and Wiley (1987) for additional information.

graphy of high relief and tectonic instability. Mount McKinley (Denali), the highest mountain in North America, is composed of graywacke and anatectic granite, which is formed at deep crustal levels from graywacke melt. Gravity and other geophysical studies support the inference that much of the 40–50 km thick crust beneath the Alaska Range is composed principally of graywacke and its more metamorphic or plutonic equivalents.

Most of the terranes in southern Alaska had been emplaced by about 60 Ma. The tectonics then, as now, involved a combination of right-slip dispersion effects along the eastern sector and subduction-related compressional effects west of the so-called armpit of Alaska at the Alaska orocline. For most of the Tertiary, subduction was not associated with additional terrane accretions. The topography became more subdued. Refraction and reflection studies have imaged subducted slabs of successively high- and low-velocity rocks that have underplated the southern margin (Figure 6.21). But beginning about 5 Ma, another large sliver, the Yakutat graywacke terrane, with possible continental basement, began accreting to the south margin of Alaska. This tectonic compressional episode has created the rugged relief that unveils both the fascination and the great complexity of this part of the world.

British Columbia

The geology of British Columbia incorporates most of the fundamental elements of terrane analysis: amalgamation, accretion, and dispersion, along with both A- and B-subduction. Recalling Figure 4.12, the western two thirds of the mountainous region of British Columbia is clearly founded on a region of suspect terranes situated outboard (oceanward) from cratonal North America. These terranes compose a collage of volcanic arcs, oceanic material, and continental fragments. Their NW-elongated geometries are oriented parallel to the continental margin. Farther east, the major part of the Rocky Mountains constitutes an east-verging fold-and-thrust belt where Paleozoic continental-margin strata have been thrust back toward the cratonal core of North America. A probable tectonic scenario that accounts for these relations follows.

The assemblage of terranes represents two composite sequences (Figure 6.23). The eastern sequence consists of upper Paleozoic oceanic material that locally has fossils diagnostic of the Tethyan zoogeographic realm. Structurally admixed with these strata are a variety of upper Paleozoic to middle Mesozoic volcanic-arc terranes, and these too have faunal assemblages that differ from coeval fossil assemblages endemic to North America at the latitude of southern Canada. The fossils are more typical of assemblages that crop out in the southern reaches of the North American continent. If one combines these faunal characteristics with paleomagnetic inclination data, which suggest approximately 10° (1100 km) of post-depositional north-directed transport, the reasonable late Paleozoic to mid Mesozoic paleogeography is found to be an Indonesian-like archipelago lying offshore from the modern region of southern California, USA, and Baja California, Mexico. For this reason this ancient volcanic-arc complex of British Columbia is often referred to as the Baja–BC arc.

Stratigraphic relations involving depositional overlaps and provenancial linkages require that this composite arc terrane was accreted to the continental margin in the area of southern British Columbia during the Late Cretaceous. Following accretion these strata became involved in the east-verging fold-and-thrust system that is commonly referred to as the

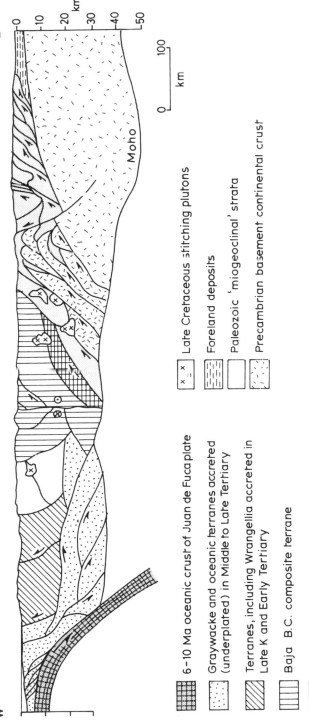

Figure 6.23 Roughly west–east crustal cross section of southern British Columbia and Alberta, Canada. Salient features include present-day B-subduction, underplating, thrust nappes of exotic terranes, and thin- and thick-skinned tectonics associated with A-subduction. (Figure from Monger *et al.* (1985))

Laramide orogeny. Two schools of thought have competed to explain this continent-directed thrusting. One suggests that the ensuing terrane accretions that were occurring along the coastal region of British Columbia pushed the more eastward Baja–BC arc terranes onto the continental margin. A more likely cause, however, invokes the idea of west-directed A-subduction that is a manifestation of ridge push in the Atlantic. This concept has already been discussed as a possible cause for the thrusting in the Brooks Range, and as is evident in Figures 4.12 and 6.20, Cretaceous cratonward thrusting characterizes the entire west margin of the North American continent, from southern Mexico to the Arctic Islands. The nature of the crustal dynamics where oceanward-directed A-subduction and continentward B-subduction are simultaneous is not fully understood; the tectonic portrayal of Timor may represent an areally restricted version of these relationships (Figure 6.5).

The western composite terrane of British Columbia consists of a Paleozoic and Mesozoic continental and oceanic fragment and several overlapping graywacke terranes that help to constrain the ages of amalgamation. The entire assemblage is stitched both internally and to the adjoining Baja–BC arc terrane by a roughly 60 Ma batholith known as the Coast Plutonic complex (see also Figure 4.10). Included in this western terrane collage are fragments of the Wrangellia terrane (see Figure 5.10), which in concert with paleomagnetic data indicate that the accretion of the western terranes involved a large component of strike-slip faulting along the western margin of North America. This coastwise transcurrent faulting, however, did not terminate after the latest Cretaceous accretion. Across the breadth of British Columbia, numerous lower and middle Tertiary pull-apart basinal sequences evince dispersion tectonics amounting to as much as 1000 km of right-slip faulting along the Straight Creek, Tintina, and associated fault systems. This kinematic pattern reflects the tectonic interactions of the North America and Kula plates (see Figure 5.7).

Sometime during the middle Tertiary, B-subduction was again initiated along the west margin as a result of North America–Farallon plate motion. This regime persists today in the region of Vancouver Island, while to the north the Pacific–North America plate motion is effecting a new episode of right-slip dispersion along the Queen Charlotte fault system. Thus, in the time frame of the past 80 m.y., a plethora of kinematic styles has shaped British Columbia. Coastwise transcurrent faulting and cratonward thrusting have dispersed composite terranes that record a complex history of amalgamation and accretion. Some of these diverse kinematic styles are coeval, reflecting distinct dynamic regimes within the same orogen. One should be neither surprised nor intimidated by such complexities.

Southern California to New Mexico

Bona fide cratonal North America lies 1500 km east of the interface between the ocean crust of the Pacific basin and the continental crust of the California Continental Borderland. The continental crust across this wide expanse is remarkably uniform in thickness, averaging about 30 km; the inferred gentle undulations in the lower crust correspond to megaboudins, that is a thinning and swelling of ductile layers caused by tensional forces. The age of the crust across this transect varies from Middle Proterozoic to Quaternary (Figure 6.24). The crust across this broad region reflects multiple episodes of tectonism and the western third of this transect consists of composite terranes – the Santa Lucia–Orocopia composite terrane and the Baja–Borderland composite terrane. For a change of pace in the manner of reconstructing the geologic history, the following presentation will begin with the most recent tectonic events and progress backward in time.

The modern physiography reflects a tectonic regime that began at 15 to 20 Ma. From eastern California to the Rio Grande rift of eastern New Mexico is the Basin-and-Range Province, a domain of east–west crustal extension, where as much as 200 km of crustal stretching has greatly thinned the crust. To the west, starting from the Salton trough and extending westward across the California Continental Border land, is a broad pliant transform-fault regime characterized by the famed San Andreas fault system. The aggregate right slip since 20 Ma for all the NW-trending faults in this region is at least 500 km. Thus, the basin-and-ridge physiography in the east part of this transect is due to extensional tectonics orthogonal to the edge of the North American continent, whereas, in the west similar-appearing basin-and-ridge physiography results from trans-pressional and transtensional tectonics tangential to the continental margin.

A limited number of paleomagnetic studies from the Baja–Borderland composite terrane (Figure 6.25) indicate that as recently as the middle Eocene, about 52 Ma, this composite terrane lay approximately 1000 km to the south. The continental nature of the Tertiary clastic sequences indicates that its place of origin lay along the margin of the North American continent in southern Mexico. Thus, the modern right-slip transform faulting along the San Andreas system is only the most recent episode of coastwise dispersion to have affected this composite terrane. Its northwesterly-directed displacements seem to have begun in the latest Eocene (Figure 6.25).

While this composite terrane was sliding along the west margin of North America, subduction-related volcanism was occurring to the north across the broad region of Arizona and New Mexico. This area was an essentially

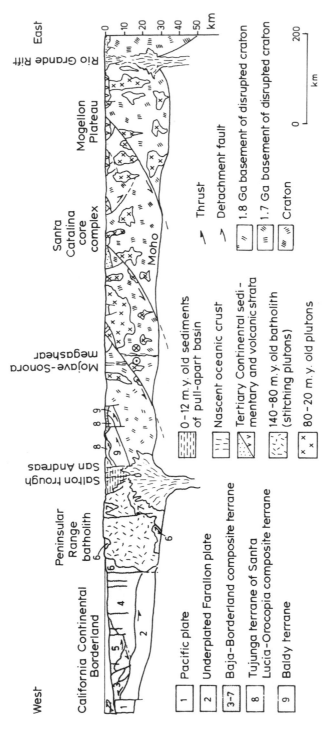

Figure 6.24 A 1500 km long west-to-east cross section from the California Continental Borderland to the Rio Grande rift. The Late Tertiary to present-day tectonics of this region are principally crustal dispersion processes: transtension in the borderland and rifting in the Salton trough, large-scale distension across the Basin-and-Range province that culminates in the mantle-tapping Rio Grande rift. These crustal extensions and rifting are superposed on an older fabric that includes thrust faults associated with accretion tectonics. (Figure modified from Howell *et al.* (1985))

flat topography, as enormous sheets of explosive volcanic material spread evenly across thousands of square kilometers. Single events represented as much as $1000 \, km^3$ of ejecta. The occurrence of this volcanism as far as 1500 km from the inferred oceanic trench suggests an anomalously flat inclination for the Wadati–Benioff subduction zone.

This volcanism was preceded by the Laramide orogeny – the episode of east-directed thrusting that affected all of western North America and which seems to reflect A-subduction stimulated by ridge-push forces from the Mid-Atlantic Ridge. Regardless of the cause, both the volcanism and the thrusting had the effect of thickening the crust. Following the relaxation of the compressional forces, the crust failed and distended in a manner reflecting the younger Basin-and-Range physiography (refer back to Figure 1.8).

Presently lying east and north of the Baja–Borderland composite terrane is the Santa Lucia–Orocopia composite terrane (Figure 6.25). In central California this composite terrane lies wholly west of the San Andreas fault. Much of the composite terrane consists of Cretaceous continental-margin volcanic arc and associated lithofacies, and, therefore, it has long been considered to be a displaced segment of an arc that originally would have connected the arc terranes of the Sierra Nevada with those of the Peninsular Ranges. But paleomagnetic data from the arc and associated lithofacies indicate that they have been displaced northward by as much as 2500 km. These paleomagnetic data are from Early and Late Cretaceous sequences. Tertiary units do not indicate anomalous paleomagnetic inclinations, and Eocene overlap sequences straddle between this terrane and autochthonous units of North America. Most of this dispersion is therefore Late Cretaceous to earliest Tertiary in age (Figure 6.26). The composition of the sedimentary detritus indicates a continental source, so as with the Baja–Borderland terrane, its transit northward has involved coastwise transcurrent faulting, presumably in a fashion similar to the more recent dispersion exemplified by the San Andreas fault system.

Recall now the previous discussion of the Baja–BC terrane of British Columbia. Paleomagnetic data from Cretaceous arc rock of that composite terrane indicate a latitude of origin that would correspond to southern California. Putting all these results together produces a Cretaceous paleogeography involving arc volcanism along much of southwestern North America, but the various arc segments have played a game of musical chairs since the time of their formation (see Figure 5.12).

The next older tectonic event that is recorded in the stratigraphic sequence along this transect is a consequence of early phases of ocean spreading in the Gulf of Mexico. Beginning in the Late Triassic, much of the eastern part of the North American continent began to stretch to presage the eventual opening of the North Atlantic Ocean. Starting at

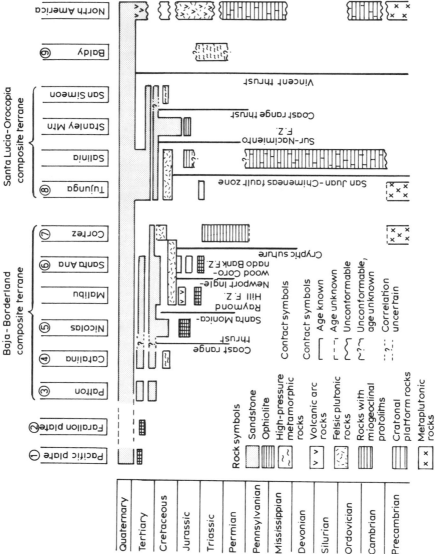

Figure 6.25 Tectonic assemblage diagram chronicalling the accretion history of terrane in southern California.

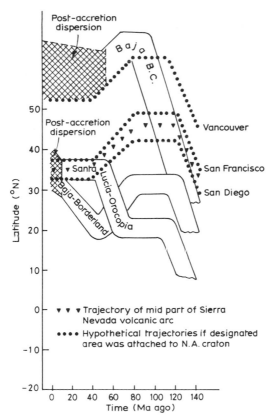

Figure 6.26 Travel paths relative to cratonal North America for Cretaceous continental-margin volcanic terranes. The hypothetical latitudinal shifts for areas along the west coast, if they had been attached to the craton, is due to clockwise rotation of the craton as a response to the opening of the North Atlantic Ocean. The estimates of latitudinal shuffling of the arc terranes is based on paleomagnetic-inclination data and reflects a long-term history of right slip between the North American continent and oceanic plates of the Pacific basin. Following the accretion of these arcs, further dispersion by faults such as the San Andreas and Tintina systems has smeared slivers of the arcs along the west coast. Only the Sierra Nevada volcanic arc has remained latitudinally fixed to the cratonal margin. (Data from Champion, Howell, and Grommé (1984), Champion, Howell, and Marshall (1986), and Irving *et al.* (1980))

about 170 Ma, ocean crust began to form in the Gulf of Mexico. Abiding by laws of plate tectonics, this new midocean ridge had to be linked into the global network of plate boundaries. The path that developed was a northwest-trending left-slip fault (see Figure 4.7) that relayed the spreading in the Gulf of Mexico to a nascent subduction zone in the eastern

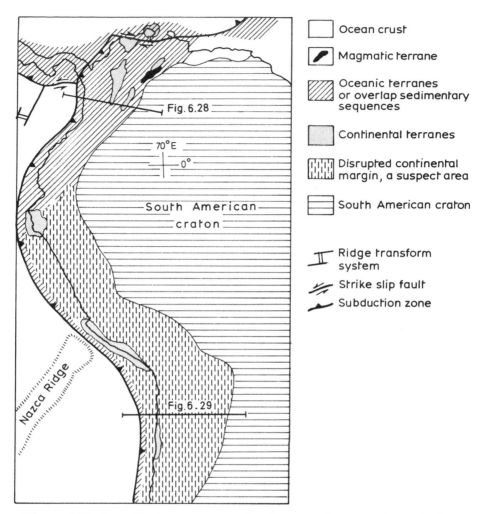

Figure 6.27 Generalized terrane map for the central and northern Andean orogen, South America.

Pacific. In southwestern Arizona and Sonora, Mexico, Precambrian terranes are offset left laterally by as much as 600 km as a consequence of this transform fault system, locally dubbed the Mojave–Sonora Megashear. Missing in this area is the Paleozoic trailing-margin sequence (miogeoclinal facies of the older literature) that is such a prominent feature of the Cordillera from Nevada to the Arctic Islands. This implies a post-Paleozoic truncation event along the southwest part of the North American continent. The Jurassic transform faulting may somehow be the culprit, but the precise kinematics causing this truncation remain obscure.

In summary, the principal post-Paleozoic tectonic events along the southwest margin of North America involve the initiation of a subduction regime as a consequence of a global reorganization of plate motions responding to the breakup of Pangea. In Middle Jurassic, left-slip transform faulting cut into the continent and moved crustal fragments toward the south into the gap being formed between North and South America. By the middle of the Cretaceous, plate motions in the Pacific possessed a component of north-directed slip relative to the North America plate and the then-young continental magmatic arcs and associated terranes were displaced to the north by varying amounts as far as British Columbia. By at least 15 Ma this coastwise dispersion was accompanied by crustal stretching that has extended 1500 km into the cratonal core. Thus, in contrast to Alaska and British Columbia, where accretion tectonics has effected a net growth to the continent, along the southwest margin there has been truncation and crustal attenuation. Most of the Cretaceous and younger material that has accreted to this margin lies to the west of the San Andreas fault system, indicating how dispersion tectonics continues today as the principal *modus operandi*; furthermore, a new ocean basin could likely form somewhere in the area of the modern Basin and Range. The basalts that are erupting through fissures in the Rio Grande rift are similar in composition to midocean-ridge basalts, indicating that primitive melts emanating from the mantle have direct access to the surface. In the restricted area of the fissures, even now there seems to be no continental crust at all. An enterprising speculator may wish to gamble on potential beach front real estate in the environs of Albuquerque, New Mexico.

6.6 THE ANDES

The Andean orogen is a 7000 km long volcanic chain that has faced an ocean basin since at least the latest Proterozoic; nonetheless, in terms of accretion tectonics, it remains enigmatic because of the spatial stability of the continental volcanic arcs and the absence of a well developed collage of accreted terranes (Figure 6.27). The history prior to the breakup of Gondwana is not well understood. The local occurrence of Paleozoic high-pressure metamorphic complexes and associated oceanic strata attests to occasional episodes of accretion, but volumetrically this material is relatively trivial. In southern Peru and western Bolivia, continental blocks with Precambrian basement are seemingly allochthonous. But whether these are rifted segments of the nearby Precambrian craton or far-traveled microcontinental fragments is not known. And other than in the region of Colombia, which is part of the Caribbean tectonic regime, evidence of accretion since the breakup of Gondwana is effectively lacking, despite almost 200 m.y. of continuous subduction along the west margin of South

America. As will be evident in the following discussion, there seem to be more questions than answers.

Colombia

In the Andean context, Colombia is out of character. The age of the volcano–plutonic complexes and the nature of accretion is more typical of the Cordillera of North America, parts of which originated not too far north of the Colombian margin. Three distinct complexes of oceanic material are well displayed in the stratal sequences, and the age of accretion is constrained by the chronology of stitching plutons and overlapping sedimentary sequences (Figure 6.28). The oceanic terranes were accreted to the Colombian margin within 10 to 30 m.y. from the time of their crystallization. Presumably this material was still relatively hot and buoyant, which must have facilitated its emplacement into the continental framework; nonetheless, the process of accretion is not well understood. In Taiwan, Tibet, the Alps, the Brooks Range, and elsewhere the obduction of ophiolitic material is always associated with the accretion of two or more thick crustal bodies, either volcanic arcs or continents. By analogy, the Late Cretaceous episode of accretion in Colombia is likely to have resulted as a

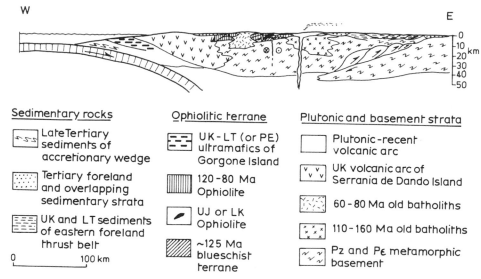

Figure 6.28 Schematic cross section through Andean orogen of Colombia. The chronology involving the accretion of oceanic terranes is well constrained by stitching plutons and successive overlap sequences. (Data from Bourgois *et al.* (1987))

consequence of the collision between the Serrania de Baudo volcanic complex and the South American continent. Furthermore, because the strata of this arc complex are rich in quartz, the basement is thought to be continental. The proposed tectonic scenario includes mid-Cretaceous continental-margin rifting, then creation of oceanic crust in the newly formed basin, followed within the next 15 m.y. by basin closure. No paleomagnetic data exist to indicate that any of the fragments are exotic, and thus a mini-Wilson cycle is a tenable hypothesis. The rifting may also explain the absence of an accretionary-arc complex associated with the Early Cretaceous blueschist strata, the oldest inferred evidence for accretion.

The youngest accretion is represented by the tectonics associated with Gorgona Island, composed entirely of mafic and ultramafic rocks, including komatiitic lava flows believed to be of Late Cretaceous or early Tertiary age. Some geologists have argued, however, that the komatiite, which elsewhere is restricted to Early Proterozoic or older stratigraphic horizons, is evidence that Gorgona Island is actually a piece of Precambrian crust. This could explain its accretion in the absence of a collision between the South American continent and another thick crustal body west of it. Until more data are collected, one must be content with some ambiguities and uncertainties regarding the nature of accretion tectonics along the west margin of Colombia.

Peru and Chile

Cropping out along the coast of Peru is the Precambrian Arequipa massif. To the east, beneath the axis of the Andes, the continental crust is between 70 and 80 km thick. Further east lies the expansive Precambrian Brazilian craton. Does this mean that a Precambrian crustal framework tracks continuously across the whole of South America? And if so, what are the tectonic processes that have thickened this crust since the breakup of Gondwana? Or alternatively, is the Arequipa massif an exotic terrane that is decoupled from the craton to the east? And if so, do other accreted terranes lie hidden beneath the surface in the thickened zone of the Andean orogen?

The crustal structure beneath the shelf regions of Peru and Chile is equally confusing. Seismic-reflection studies indicate that continental crystalline basement extends westward to within 20 to 30 km of the modern trench axis (Figure 6.29). The volume of accreted sediment is menacingly small for accretionary aficionados. Mass wasting and normal faulting are prominent features displayed by the stratigraphy and structure of Miocene and younger units. Furthermore, recent borehole stratigraphic data gathered by the Ocean Drilling Program, Leg 112, indicate that the midslope region is characterized by tectonic collapse in front of a gently

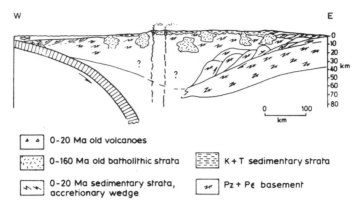

Figure 6.29 Generalized E–W cross section through the Andean margin of South America at the latitude of southern Peru and northern Chile.

flexed upper slope. A flattening of the subduction zone may have created a supercritical taper (Figure 2.13), which leads to extensional faulting. Thus, rather than continental growth concomitant with subduction and accretion, the margin is thinning and collapsing. This has also prompted some geologists to invoke a concept called *tectonic erosion*, that is, the milling away of the continental margin. These geologists have suggested that the rough surface of the subducting oceanic plate is like a buzz saw. With this idea in mind one might ask if the continental sawdust is underplating the margin to the east producing some of the thickening of continental crust beneath the Andean axis?

In summary, at least three processes may be said to contribute to the thickening of the Andean orogen. East-verging thrusts along the east side of the range indicate an A–subduction phenomenon, presumably caused by ridge-push from the Atlantic, similar to the Laramide orogenic features discussed for the Cordillera of North America and to the thrusting in the Brooks Range. Thrust nappes, particularly if the Precambrian basement is involved, would thicken the crust. In addition, batholithic material has been invading the crust in a discontinuous manner since the Precambrian. In parts of southern Chile, the volcanic strata since 200 Ma display a characteristic geochemical evolution in their Rb/Sr systematics. The initial ratios of $^{87}Sr/^{86}Sr$ have progressed from 0.708 to 0.702. This indicates an enhanced dilution of continental material by primitive mantle melts. And finally, underplating and associated deep-crustal accretionary events could also be thickening the crust. The absence of a diverse assemblage of mappable suspect terranes may compromise the last suggested process, but until more deep-crustal data are available the full explanation for the Andes will remain a mystery.

6.7 CONCLUSION

No simple tectonic model is applicable to all active continental margins. Nonetheless, from this brief survey of the tectonic regimes, from the archipelagos of the southwest Pacific, through the Tethyan systems of southern Eurasia, to the Cordilleran and Andean systems of western North and South America, one should appreciate the inherent mobility of continental crust. This crust has grown since 200 Ma, a continuation of a process that started at 4 Ga. The resiliency of continental crust is due to its buoyancy. Subduction volcanism creates new continental crust. Relative to oceanic crust and its lithosphere, continental material is weak. Shear along ocean–continent interfaces results in faulting that follows a continental pathway, and, therefore, plate-tectonic processes are continuously moving continental fragments from place to place. In collision zones the crust thickens, thereby forming the roots to major mountain chains. Because of its inherent weakness, compressive forces within continental crust also cause it to yield along low-angle thrust planes and this results in the crust becoming thicker by a process called A-subduction. But if thickened too much, the crust will distend and spread laterally. Determining the rates, timing, and sense of movement for all these processes is both the challenge and the joy of field geology.

7

The strategy of a field geologist

7.1 THE ROLE OF FIELD GEOLOGY

Numerous national, international, and provincial geologic societies around the world have either just celebrated, or are about to celebrate, their centennial anniversaries. The founders of these organizations and the many members that followed them were lured into geology by a combination of curiosity, a love of the outdoors, a quest for exploration, a need to climb the mountain, and a longing to conquer the last frontiers. The process was as important as the product. The aesthetics of the outdoors were inextricably entwined with analytical exercises. Great expeditions were launched across the Alps, the Himalayas, and the Rockies. Geologic maps were the tablets that recorded their discoveries. Vast collections of rocks and fossils were sent back to laboratories and museums for further study, and the explorers kept pushing forward, shrinking the frontiers.

The first half of the 20th century was an extraordinary era of geologic mapping that was interrupted only briefly by the two world wars. Stimulated by a quest for mineral and petroleum resources, geologists were literally roaming the globe. The life of petroleum or mineral explorationists was one of constant travel; from the tropical highland jungles of New Guinea to the frigid tundra of the Arctic islands, geologists were mapping geologic structures, measuring stratigraphic sections, and collecting rocks and fossils. Momentous discoveries were made. By the early 1960s, when the plate-tectonics paradigm was being developed, geologists had already located most of the world's major mineral deposits and petroleum reservoirs. Since then, only a few remote or inhospitable localities, such as the North Sea, the far offshore Gulf of Mexico, and the Arctic Prudhoe Bay in northern Alaska, have added to the inventory of major oil reserves. Cartels and political unrest have created short-term instabilities in the price of natural resources, but on average, the price of oil has not kept up with the rate of inflation for the past three decades. The price of copper and iron remains at 1950s valuation. The financial attraction of exploring for natural resources is no longer what it once was; consequently, the financial backing for onland geologic exploration and mapping has been greatly reduced.

The reduction in resource exploration is not the only factor contributing to the scaled-down effort toward geologic mapping. As recently as the early 1960s, a commendable university thesis might have focused on a quadrangle mapping exercise. But as the plate-tectonic paradigm began to capture the imagination of the geologic community, descriptions of local structures or stratigraphic relations were no longer stylish. Theses now need to seek solutions to well defined scientific problems. Geologic models have to be proposed and tested. New emphasis is being placed on narrowly focused scientific problems at the expense of exploratory geologic mapping. University and research-laboratory promotion panels give scant praise to a map; instead, evaluations concentrate on the volume and quality of a person's *Vitae*, hopefully replete with 'refereed' journal articles. Junior scientists are compared by their potential for a scientific breakthrough, and rarely on their mapping skills. Yet, the geologic map is the foundation upon which most models are built. Without a map, where are you? One may liken the lack of a geologic map for a modeler or theoretician to the absence of a road map to a driver in an unfamiliar area – or to the lack of a blueprint for a housebuilder (imagine the architecture of the completed structure without one).

The past several decades have also been an era of specialization. A dizzying array of new tools and technologies have replaced geologists' boots and Brunton compasses: photogrammetry, scanning–electron microscopy, acoustic tomography, magnetotellurics, multichannel seismic reflection, and cryogenic magnetometers, to name a few.

Research monies are provided for short-term projects, usually covering only one or two years, hardly enough time to conduct thorough research mapping. Funding agencies are looking for focused experiments or narrowly defined field investigations. Big, coordinated efforts, involving teams of scientists, encompassing a wide range of expertise, are more in vogue. But here too the focus is relatively narrow. The almost random yet magnificently fruitful explorations of the Deep Sea Drilling Program of the 60s and 70s have been replaced by thematically focused investigations of the Ocean Drilling Project. Regional mapping is being superseded by corridor or transect programs where geologists and geophysicists concentrate their investigations along a narrow cross section of a foldbelt. Deep continental drilling is replacing regional field studies in an effort to find and sample strata representative of deep crustal horizons.

If some of the luster of field mapping has been tarnished by the difficulties of acquiring funding or by the lack of recognition of the importance of geologic maps, the lure of field-centered projects still exists. And just as definitely, field geology still has a vital role to play.

In the oceanic domain, the precepts of rigid-lithospheric plate tectonics are essentially proven hypotheses. The application of the plate-tectonic paradigm to the continental domain, however, remains clouded by uncertainty. The very name of 'suspect' terranes conveys this uncertainty, and an example of applying plate tectonics to a specific feature is particularly illustrative. At numerous localities throughout the Coast Ranges of California one can observe a sequence of strata that is made up principally of graywacke, locally including basalt, chert, or high-grade metamorphic blocks. These strata are highly tectonized, and in places they are overprinted by regional blueschist or greenschist metamorphism. Originally called the Franciscan Formation, this sequence is now variously referred to as the Franciscan assemblage, melange, complex, or terrane. Until the mid-1960s these strata were generally regarded to lie unconformably below a thick sequence of Upper Jurassic to Upper Cretaceous arkosic sandstone and shale, called by a number of formation names, but best known as the Great Valley sequence. Careful mapping, however, demonstrated that a discontinuous sequence comprising all or part of the following: chert, basalt, diabase, gabbro, serpentinite, and ultramafic rock, is often sandwiched between the arkosic Great Valley sequence and the Franciscan assemblage. Geologists familiar with plate tectonics suggested that these chert to ultramafic units, lumped into the Franciscan assemblage by early workers, represented the disrupted ophiolitic basement of the Great Valley sequence, and the contact between them and the Franciscan assemblage was a fault, which became known as the Coast Range thrust.

The Coast Range thrust became a keystone of the early models for plate-tectonic control of the geologic architecture along convergent continental margins. At first, the Coast Range thrust was viewed as a fossil subduction zone. Implicit in this notion was a genetic linkage between the Franciscan units, the Great Valley sequence, and the Jurassic and Cretaceous volcanic rocks of the Sierra Nevada – respectively the accretionary prism, forearc basin, and volcanic arc. But some geologists contested this linkage, based on inferred compositional differences in the respective source terranes of the Franciscan and Great Valley sequences. For these geologists the Coast Range thrust was viewed as a tectonic suture. Most recently, coincident with the recognition of the importance of extensional faulting in collision zones, the characterization of the Coast Range thrust is changing again. This fault is now interpreted to be a young detachment surface, that is, a low-angle normal fault. Clearly our vision suffers from myopia; objectivity is always tainted by available concepts.

Which tectonic model best portrays the geologic history of the Coast

Range 'fault' is unknown. In fact, the kinematic history of the fault may be so complex that aspects of all of the proposed models (except the unconformity) are locally recorded along its trace. The best resolution of these controversies will be established only by a combination of more mapping and more analysis. Similar unresolved questions exist for most orogenic foldbelts. Although plate tectonics has provided us with a framework – a rationale by which to analyze tectonic processes – we have not yet reached the stage where our understanding of continental geology is comparable to our understanding of the geology of oceanic plates. More mapping is required to define the geometric relations better. Kinematic fingerprints must be sought within selected areas. Many rocks and strata remain undated, and the composition of stratal sequences remains poorly defined. Without these critical data we will not be able confidently to describe the Earth's geometric relationships, the kinematics of terranes, and the dynamics of continental plate-tectonic processes.

Scientific journals have burgeoned in this era, and the number of geologic articles published worldwide each year has reached almost overwhelming proportions. In the early 1970s, the yearly production of geologic papers was approximately 700 000, and by the middle 1980s this had increased to 1 500 000 papers. Yet during the same time, the number of published geologic maps has fallen dramatically. And the fate of field geology is possibly symbolized by this decline. Since 1980, the yearly publication of standard geologic quadrangle maps within the US Geological Survey has averaged about eight; whereas in the 1970s the Survey published an average of 50 to 60 geologic maps per year. Cost is often cited as a reason for decreased map publication, but if throwaway posters can be produced in abundance for every imaginable rock'n'roll singing group, surely the geologic community can find a way to produce and distribute low-cost maps. Modern maps are syntheses of the new tectonics, catalogues of data, and inventories of current knowledge. Geologic maps are progress reports, cornerstones of scientific methods, and the real point of contact between ideas and reality.

7.2 A FIELD-MAPPING STRATEGY

The development of new technologies and forced specialization require a team approach to most field investigations. The consummate field geologist must coordinate these efforts, identify the problems, and select the disciplines and the investigators who can offer the necessary solutions. In many respects, a good field party can be likened to a well fitted research vessel, one that is capable of mapping the seafloor, collecting rock samples for paleontologic and geochemical analysis, imaging subsea reflectors, characterizing paleomagnetic signatures, measuring gravitational and magnetic fields, performing state-of-stress measurements, and extrapolating these data into a regional framework.

Who are the field geologists? What would be a typical professional profile of a representative individual? In essence every geologist has the potential to fill this role. To be effective, however, a number of personal traits and professional qualifications v/ould be desirable. The individual should command a broad working knowledge of many disciplines. Though possessing a particular expertise is inevitable in this period of specialization, one must be able to communicate with the broad spectrum of specialists representing the vast array of disciplines that will necessarily play a role in any thorough field study. Some field geologists may concentrate most of their professional lives on a particular region, a specific orogen, or a well defined stratigraphic horizon. An intimate knowledge of every outcrop suits the personality of some individuals, while others may be equally comfortable venturing into new territories, analyzing any part of the stratigraphic record. Regardless of their professional and personal persuasions, field geologists must have a 'nose' for the *significant* problem. The field geologist must stay abreast of current concepts regarding Earth processes. Extrapolation in the third dimension is as important an asset as the facility to order salient features chronologically. Most geologists like to think of themselves as possessing these characteristics, although the 'generalist' appellation may be a negative factor in some circles. This label often carries with it a pejorative connotation when it comes to securing a research grant, finding a job, or getting promoted. Being a jack of all trades but a master of none is not suitable for everyone, nor is it desirable for the profession as a whole. But just as with medicine, where the desirability of the 'family doctor' or general practitioner is again being recognized, geology needs a critical mass of generalists to help direct and focus the specialists to the important regional problems, to the critical outcrops.

Funding and organizational frameworks must be flexible enough to allow for investigational fluidity. Data need to be shared in an open arena by geochemists, paleontologists, petrologists, geophysicists, and tectonicists. The field geologist can play the role of guide, broker, and diagnostician, akin to the chief scientist on a research vessel. Specialists often need and welcome direction and coordination. The field geologist should be as much concerned about problems as solutions.

Organizing a research team

The solution to most present-day problems requires a team approach. For problems in tectonics, an entire entourage of geologists may be required, often working independently, but coordinated in some functional way. Geologists must stay abreast of the development of new tools and techniques. For example, satellite and various kinds of airborne imagery are important tools now available to land geologists. Similarly, marine geologists now have at their disposal a spectrum of tools that image the seafloor with resolutions comparable to those ranging from satellites to low-flying airplanes. These 'photo mosaics' enable one to detect local and regional patterns that facilitate spatial extrapolation.

Dating rocks and minerals and recording the chronology of events is critical to all investigations. Almost every imaginable species of plant or animal, from dinosaurs to algae, housed in strata composed of shale, carbonate, or chert, are potential aids in a field investigation. Consequently, numerous paleontologists may have to be called upon. Geochemistry, involving the various radiogenic decay systems, such as U to Pb, K to Ar, Sm to Nb, or Rb to Sr, provides opportunities to date crystallization events. Isotopic systems controlled by ancient climatic and tectonic processes, for example variations in $^{87}Sr/^{86}Sr$ ratios, are also useful tools for dating ancient marine sediments. Thus, numerous investigators, utilizing a vast array of analytical procedures, are necessary players in the dating efforts of a field investigation. Variations in blocking temperatures within a suite of minerals offer a way to determine tectonic uplift rates. Past thermal regimes can be inferred from metamorphic facies analyses as well as by the maturation level of organic compounds. Even variation in the color of some fossils, such as conodonts, indicates past ambient thermal histories.

In the 'good old days', cross sections were drawn based on a geologist's ability to visualize and extrapolate surface features to deeper crustal levels. A variety of modern techniques eliminates much of the guesswork. Seismic reflection, seismic refraction, and magnetotellurics all provide vital constraints on the subsurface geometry of geologic features. These subsurface profiles can be an invaluable aid to a field investigation, because they direct the field geologist to the most significant features.

Other forms of geophysical modeling involving potential-field data, such as magnetic and gravity, also provide insight into subsurface relationships. For example, the thickness of a sedimentary basin or the composition of the roots of a mountain system may be modeled. Paleomagnetic investigations should also be performed on a routine basis in all areas to search for terranes that are rotated or tectonically displaced.

The compositions of rocks and strata are critical ingredients to all field investigations. Provenancial linkages, depositional environments, paleoclimatic conditions, and post-depositional alterations are just a few of the important parameters read from sedimentary rocks. Environments of emplacement of igneous rocks are also discriminated on the basis of variations in rock, mineral, and elemental compositions.

The aforementioned is a sampling of the arsenal of tools and disciplines upon which the field investigator can draw in the planning and execution of research. The age of specialization and diversification has rung the death knell of the natural scientist, and no individual can master or perform all the necessary tasks and no funding agency will support all the investigators necessary to answer all the questions. But a cunning field geologist can identify the most important problems and by one means or another attract the right people and sufficient money to solve the multitude of geometric, kinematic, and dynamic problems that remain unresolved.

7.3 CONCLUSION

To one degree or another, all of the Earth's surface has been mapped geologically. The plate-tectonic paradigm not only provides a rationale to integrate Earth processes, but the very nature of plate tectonics requires a kinematic and dynamic linkage among all of the Earth's plates as well as an integration of core, mantle, and crustal phenomena. In this book, I have tried to highlight some of the major conclusions as well as some of the fundamental problems that exist within our current state of knowledge. On the one hand, because of technological advances, geology has entered into an era of specialization, yet because of the interconnection of diverse geologic phenomena, any explanation must integrate a plethora of data and a host of factors many of which may seem quite unrelated. India crashing into Asia may have affected the kinematic behavior of the crust in the North Atlantic Ocean. Deep-mantle plumes may be influencing the isotopic chemistry of surficial volcanic rocks. The rate of midocean spreading can determine continental freeboard. Because plate-tectonic processes have probably been operating for most if not all of Earth history, the spatial relations among crustal masses must be considered suspect, and independent data are needed to constrain the kinematic history of each stratigraphic sequence.

For most of the past century, the role of the field geologist was to map and characterize the distribution of rocks that crop out on the Earth's surface. In the process of this demanding task, most of the extractable resources were located and described. This era of mapping is essentially complete. A new era must begin, one where the objectives are to describe in greater detail the nature of the crust and to describe the vexing kinematic history of each segment. Data must be rigorously recorded on maps, where each line and pattern is fully documented, extrapolations are minimized, and supporting data are referenced. Contacts between stratal units must not only be characterized, but the kinematic history along these contacts must be investigated in a way that clarifies the different styles of motion that may have occurred. Vertical and horizontal movements must be integrated with global fluctuations of sea level. The time of crustal events must be calibrated in order to discriminate between local and global disturbances.

In the context of how geologists understood the workings of the Earth just 30 years ago, it may seem that most problems have been solved. In pre-plate-tectonic days, the ocean was understood to be foundered Pre-cambrian crust; now not only do we know it to be young material that is recycled on average every 110 m.y., but geologists have mapped the strati-graphic distribution of the oceanic crust on a global scale. Foldbelts are no longer considered to be geosynclinal features resulting from vertical perturbations, but rather features that cover a spectrum of plate motions from sliding, to colliding, to rifting. Nevertheless, a host of unsolved questions remain. What is the history of continental growth? How much, if any, continental crust is recycled into the mantle? How did the Andes form? How do core and mantle dynamics affect the motion of lithospheric plates? What causes continental, A-type subduction? What was the paleogeography of the Neo- and Paleo-Tethys? Where were the origins of the continental terranes of Alaska?

The above questions are only a sampling of much that remains unknown. The book began with a few questions, some of which I hope to have answered. But each question generates more questions, and one is never certain if one is getting closer to the ultimate solution or not. Questions at every scale abound, and the form of these questions will continue to evolve as we sharpen our tools and our wits. For those who wish a lifetime of being carried intellectually and physically across our globe – the field still beckons.

Glossary of terms

The following list of terms includes abbreviated definitions germane to the context of this book. For more elaborate or inclusive definitions one is advised to consult appropriate text books or any of the published glossaries of geology.

Accretion In terms of terrane analysis, the structural joining of two or more allochthonous bodies; in a more restricted sense, it refers to the tectonic incorporation of terranes into a continental framework. In terms of subduction, the process of scraping sediment off a subducting slab and incorporating it into an accretionary prism.

Accretionary prism A generally wedge-shaped mass of tectonically deformed sediment, possibly with minor components of ophiolitic material, formed in a subduction zone as a result of the tectonic transfer of strata from the descending plate into the framework of the overlying plate. The base of an accretionary prism is a decollement and its leading edge is the deformation front. Also *accretionary wedge*.

Asthenosphere A region within the Earth's mantle, below the 1400°C isotherm, where the crust behaves plastically and over which the more brittle lithosphere can move.

Allochthon A crustal fragment that has been displaced.

Allochthoneity The extent to which something has moved either from its place of origin or from the last place that it was known to have resided.

Allochthonous Adjective describing something that has moved, usually in reference to the continent on which it formed or the continent to which it accreted.

Amalgamation In terms of terrane analysis, the same as accretion, but implying the joining together of two or more bodies in an oceanic setting away from a continental margin.

Anatexis The melting of pre-existing rock, particularly quartz-rich sediments that are exposed to extreme heat by deep tectonic burial. Adjective: *anatectic*.

Apparent polar wander The apparent wander of a magnetic pole assuming a fixed position for a continent; in fact, the magnetic pole is nearly fixed, corresponding to the pole of the Earth's spin axis, and the continents' wander.

Aseismic ridge An oceanic plateau that does not display any present-day volcanism or seismicity, for example a continental fragment, remnant arc, and fossil hotspot edifice, as opposed to active midocean spreading ridges, volcanic arcs, and hotspot volcanoes.

Backarc basin A basin floored by oceanic crust that formed about coevally with the volcanic arc and that is positioned on the opposite (back) side of the arc from the subduction zone.

Belt In a tectonic sense, a general term for a distinct assemblage of rocks that are aligned in a linear pattern.

Blind thrust A thrust plane that does not reach the Earth's surface, commonly inferred from the folding of rock layers presumed to lie above such a thrust.

Blocking temperature In reference to paleomagnetism, the temperature below which magnetic orientations become frozen in space; in reference to mineral crystallography and radiogenic dating, the temperature below which the products of radioactive decay are retained in the crystal lattice.

Bolide A general term for any foreign body that collides with the Earth.

Bouillabaisse A stew-like soup made up of a variety of fish and shellfish products, and sometimes other things as well.

Collage A patchwork map pattern of accreted terranes.

Composite terrane A tectonostratigraphic terrane, made up of two or more terranes associated by a common geologic history prior to accretion at their location.

Conjugate margins The margins on opposite sides of an ocean or basin that were joined prior to rifting.

Consolidation The process whereby mobile foldbelts are gradually transformed into rigid cratons, the so-called continentalization of orogenic belts.

Continental freeboard Mean elevation of land above sea level.

Contractionist One who explains orogenic foldbelts as being due to a shrinking Earth as a consequence of long-term cooling.

Craton The part of continental crust that has been stable for a long period of time, generally at least since the Proterozoic.

Cryptic A feature hidden from view, said of terrane boundaries that may be obscured owing to a metamorphic overprint, an engulfing pluton, or simply the lack of outcrop.

Decollement A surface of detachment, generally a subhorizontal plane across which rock bodies either move or are deformed in independent styles.

Deformation front The initial point of contact between two plates in a subduction zone; the leading edge of an accretionary prism.

Dispersion Tectonic disruption that results in large-scale separation of crustal bodies.

Drag forces Frictional forces along the interface between the viscous asthenosphere and lithosphere, commonly directed against the gravitational body forces that drive the motion of the lithospheric plates.

Dynamic appraisal An analysis of the forces responsible for the kinematics of geologic features.

Duplex A sequence of thrust sheets where individual sheets are bounded on top and bottom by thrusts.

Eugeosynclinal deposits An archaic term for graywacke and chert-rich sedimentary and andesitic to basaltic strata that are typically found in orogenic belts.

Euler pole The pole of rotation describing the motion of a rigid plate across the spherical surface of the Earth.

Exotic An allochthonous body that originally formed in a foreign setting.

Fixism A belief in the spatial permanence of continents and oceans.

Flake A tectonostratigraphic terrane in the form of a thrust nappe, formed during continent–continent collision.

Flysch basin A basin where the principal deposits are thick sequences of submarine-fan turbidite deposits.

Ga. Giga-annum, 10^9 years.

Geocentric dipole A model of the Earth's magnetic field such that approximately 90% of the magnetic forces can be explained as if a bar-magnetic dipole is oriented coincident with the Earth's spin axis.

Geometric appraisal An analysis of the spatial relations of geologic features.

Geosyncline An archaic term used by oscillationists to suggest a mobile downwarping of the Earth's crust before orogenesis. A plethora of geosynclinal subdivisions were invented in an attempt to explain the complex array of geologic relations along continental margins. Plate tectonics has supplanted geosynclinal theory as the working paradigm.

Gondwanaland From Gondwana, 'Country of the Gonds' a region of central India; Gondwanaland is the continent including the country of the Gonds consisting of the agglomeration of landmasses in the southern hemisphere that existed in the late Paleozoic (South America, Africa, India, Australia and Antarctica). The accretion in the latest Paleozoic of Gondwanaland to the northern hemisphere continental agglomerations resulted in the formation of the supercontinent Pangea. Here and elsewhere Gondwanaland is often shortened to Gondwana.

Gravitational body forces The forces resulting from the pull of gravity upon the lithosphere as opposed to driving forces associated with a coupling between asthenospheric convectional flow and the base of the lithosphere.

Greenstone belt In Precambrian terranes, the assemblages of volcanic strata (New Zealand geologists use this term to denote belts of serpentinite).

Hydrostratically overpressured The situation which results when the internal fluid pressure at a given depth exceeds the pressure due to the weight of a column of fluid reaching the Earth's surface. When this pressure is equal to the confining lithostatic pressure the strata above the overpressured horizon are effectively 'floating'.

Island arc The arc-shaped row of volcanoes above a subduction zone, for example the Aleutian or Mariana arc.

Isostasy A condition where low-density crustal rock floats in a viscous mantle.

Isotherm Surface of equal temperature.

Kinematic appraisal Judging the movement history of geologic features based on geometric relationships.

Lacuna A gap, hiatus.

Lithosphere The 50–150 km thick plates that characterize the outer sphere of the Earth. The relatively brittle lithospheric plates ride above the plastic asthenosphere. Includes crust and uppermost mantle.

Ma. Mega-annum, 10^6 years.

Magnetic inclination The vertical component of the magnetic field at any point of the Earth's surface. At the magnetic equator the inclination is zero and this angle increases progressively, reaching 90° at the magnetic pole.

Magnetic lineations The magnetic stripes produced in oceanic crust by alternating episodes of magnetic reverse and normal polarities, forming lineations that are generally symmetric about midocean spreading ridges.

Magnetic declination The horizontal component of the magnetic field at any point on the Earth's surface; the angle of deviation from the Earth's spin axis which for a modern reading is due to precessional wander of the geocentric dipole and for paleomagnetic readings may be due to tectonic rotations.

Magnetostratigraphy Stratigraphic correlations utilizing the super-positional record of reverse and normal polarities of the Earth's magnetic field.

Massif A massive structural body cropping out in an orogenic belt.

Melange A descriptive term for a chaotic assemblage of rock types, commonly with a fabric of randomly oriented blocks in a finer-grained matrix of either mudstone or serpentine. Their occurrence may reflect mixing due to tectonism, diapirism, or mass-wasting.

Microplate A small lithospheric plate.

Miogeosynclinal deposits An archaic term for the thick, nonvolcanic, carbonate- and clastic-rich strata along a continental margin.

Mobile basement A nappe consisting of deformed basement rock generally considered to represent rock from the ductile zone of the lower crust.

Mobilism A recognition of the potential for significant mobility of the Earth's continental and oceanic crust.

Nappe A thrust sheet of a lithologic assemblage distinct from the adjoining rock units.

Obduction A general term to describe the tectonic emplacement of oceanic material above continental crust.

Offscraping Tectonic transfer, at the deformation front of a subduction zone, of sediment from the lower-descending plate to the upper, relatively stable plate.

Oceanic plateau A topographically elevated region of the ocean floor where the crust is generally 15–40 km thick as opposed to the 6–8 km thickness of normal oceanic crust.

Ophiolite A sequence of igneous rock that includes basalt, diabasic dikes, gabbro, and peridotite thought to represent the oceanic crust.

Oscillationist One who ascribes orogenic foldbelts, continents, and oceans to thermally-driven vertical tectonics and attendant gravitational sliding.

Overlap sequence A depositional sequence that unconformably lies across the contacts of two or more terranes, sometimes referred to as a successor basin sequence.

Paleomagnetic data Ancient magnetic-field orientations recorded in a rock.

Pangea Sometimes spelled Pangaea. The supercontinent that existed 300 to 200 m.y. ago consisting of a configuration of nearly all continental crust. Pangea is the hallmark of Alfred Wegener's theory of continental drift.

Panthalassa The now-vanished ocean that surrounded the supercontinent of Pangea.

Paradigm A model, conceptual framework that explains a set of observations.

Parautochthon A crustal block that has moved a relatively short distance within the domain of the continent on which it is now located.

Parautochthonous Adjective describing something that has moved, but only a relatively limited amount, especially in reference to the continent on which it formed or on which it was last emplaced.

Piercing point The point of intersection of a geologically defined line or point with a fault plane; recognition of offset piercing points allows the precise determination of amount and sense of displacement of the fault.

Precession A systematic change in orientation of the axis of a spinning sphere due to torque.

Provenancial linkage Correlation of two or more terranes based on the composition of the sedimentary units; a tectonic linkage among terranes is inferred if one or more terranes was the source of the detritus deposited on one or more adjoining terranes.

Province A general term for a broad region characterized by either a particular structural style, distinct stratigraphy, or unique geophysical signature or combination of these characteristics.

Pull-apart basin Basin that forms in a regime of crustal thinning and local rifting as a result of transtensional faulting.

Pure shear The flattening of a body due only to elongation.

Remnant arc An extinct volcanic ridge that has been abandoned as the active region of volcanism, most commonly as a consequence of backarc spreading or a jump in the location of subduction.

Remanent magnetization The magnetic moment within a rock after the removal of an applied magnetic field (the latter is the *induced* magnetization); the remanent magnetization may be made up of a variety of inherited magnetic orientations. The objective of magnetic cleaning is to remove younger superposed orientations in an attempt to isolate the primary remanence.

Ridge push The gravitational body forces of a lithospheric plate sliding away from a spreading ridge. The inclined slope is defined by the approximate location of the 1400°C isotherm.

Rigid basement A nappe consisting of undeformed crystalline basement generally implying an origin within the upper 10 km brittle zone of the crust.

Seismic reflection A technique where sound (acoustic waves), at varying frequencies, is sent into the Earth and the record of returning signals (sound that has bounced or reflected) provides a view of subsurface geologic features. The lower the frequency of the sound source the greater the penetration and the less the resolution.

Seismic refraction A technique of seismic shooting where numerous stations record the arrival times of refracted sound waves from a single source, commonly a high-energy explosion. These records may provide information regarding the configuration of high-velocity (dense) layers in the middle and lower portions of the crust.

Sialic crust Continental crust where rocks contain minerals rich in silica and alumina, for example granite with quartz and orthoclase feldspar.

Simatic crust Oceanic crust or the lower portions of continental crust, where rocks contain minerals rich in silica and magnesia, for example gabbro and basalt with olivine and pyroxene.

Simple shear The parallel movement of one plane relative to another effecting a rotation of fabric elements.

Slab pull The gravitational body forces associated with the descent of a lithospheric plate into a subduction zone.

Stitching pluton A plutonic body that intrudes a suture zone or fault contact between two terranes.

Strain The change in shape or volume of a body of rock as a result of stress.

Stress Force acting on a body of rock.

Subduction The process where one crustal lithospheric plate descends beneath another. A term originated by Alpine geologists in the pre-plate tectonic era to describe large-scale thrusting. A-subduction is wholly within continental crust, whereas B-subduction involves oceanic crust descending beneath either continental or oceanic crust. Subduction does not necessarily imply the transfer of rock into the asthenospheric part of the mantle; and indeed, an important hypothesis incorporated in the thesis of this book is that most continental crustal strata cannot be subducted into the mantle owing to the relatively low density of the crust. The buoyancy of continental crust keeps it 'floating' above the mantle.

Subduction erosion The tectonic removal of strata from the leading portion of the upper lithospheric plate in a subduction zone.

Suture zone The tectonic expression of a collision zone, commonly consisting of a zone of ophiolitic or high-pressure metamorphic material. Because the current distribution of many tectonostratigraphic terranes is due to post-accretion dispersion, many terrane-bounding faults are not suture zones.

Tectonic assemblage diagram A form of a correlation diagram that characterizes the history of terrane accretions and amalgamations.

Tectonostratigraphic terrane A fault-bounded package of strata that is allochthonous to, and has a geologic history distinct from, the adjoining geologic units.

Terrain The lie of the land; physiographic and topographic features.

Terrane A geologic entity. The term requires a modifier, for example, granitic terrane, Precambrian terrane; in this book, all terranes are tectonostratigraphic terranes.

Thick-skinned Adjective describing thrust nappes that include basement rock.

Thin-skinned Adjective describing thrust nappes consisting only of the strata detached from crystalline basement.

Transform fault A plate-bounding, strike-slip fault that connects ridges, trenches, or other transform faults or some combination thereof.

Transcurrent fault A large-scale strike-slip fault.

Transpressional Adjective describing strain consisting both of lateral and contractional motion, typical of zones of uplift along strike-slip faults.

Transtensional Adjective describing strain consisting of both lateral and extensional motions, typically resulting in the formation of strike-slip pull-apart basins.

Trench rollback The progressive migration of the trench axis of a subduction zone in a direction toward the descending plate, that is, the absolute motion of the arc, forearc, and trench is forward, in the direction opposite to the subducting plate.

Trench suction A force applied to the upper plate of a subduction zone as a result of trench rollback and the flow of asthenospheric material into the region beneath the leading part of the upper plate.

Underplating The tectonic transfer of material to the underside of an accretionary prism or volcanic edifice in a subduction zone; sometimes also called subcretion and contrasted with accretion, where the tectonic transfer of material occurs along the deformation front.

Uniformitarianism A dialectic whereby modern geologic processes are studied in order to explain ancient geologic relations.

Wadati–Benioff zone The locus of deep earthquakes that corresponds to the plane of subduction, named in honor of the two independent discoverers.

Wilson cycle An accordion style of plate tectonics involving the opening and closing of ocean basins, named in honor of J. Tuzo Wilson.

References

The following references are recommended reading. The list includes citations mentioned in the various illustrations of the book as well as articles not specifically cited but which may be of general use to the interested reader.

Allegre, C.J. (1982) Chemical geodynamics. *Tectonophysics* **81**, 109–132.

Alvarez, W., Kent, D.V., Premoli-Silva, I., and Schweikert, R.A. (1980) Franciscan complex limestone deposited 17° south paleolatitude. *Geol. Soc. Amer. Bull.* **91**, 476–484.

Armijo, R., Tapponnier, P., Mercier, J.L., and Han T.-L. (1986) Quaternary extension in southern Tibet; field observation and tectonic implications. *J. Geophys. Res.* **91**, 13 803–13 872.

Armstrong, R.L. (1981) Radiogenic isotopes; the case for crustal recycling on a near-steady-state no-continental-growth Earth. *Philos. Trans. R. Soc. London Ser. A* **301**, 443–472.

Arrhenius, G. (1985) Constraints on early atmosphere from planetary accretion processes. Lunar and Planetary Sciences Institute Rep. 85-01, 4–7.

Atwater, T. (1970) Implications of plate tectonics for the Cenozoic tectonic evolution of western North America. *Geol. Soc. Amer. Bull.* **81**, 3513–3536.

Aubouin, J. (1965) *Geosynclines*, Elsevier Publishing Company, Amsterdam.

Aubouin, J. (1984) Méditerranéenne (Aire). *Encyclopedia Universalis*, 2nd edn, vol. 11, Paris, France.

Aubouin, J., LePichon, X., and Monin, A.S. (eds) (1986) *Evolution of the Tethys. Tectonophysics* **123**, 1–315.

Bally, A.W., Gordy, P.L., and Stewart, G.A. (1966) Structure, seismic data and orogenic evolution of southern Canadian Rocky Mountains. *Bull. Canad. Petrol. Geol.* **14**, 376–395.

Beck, M.E. Jr (1980) Paleomagnetic record of plate-margin tectonic processes along the western edge of North America. *J. Geophys. Res.* **85**, 7115–7131.

Bishop, D.G., Bradshaw, J.D., and Landis, C.A. (1985) Provisional terrane map of South Island, New Zealand, in *Tectonostratigraphic Terranes of the Circum-Pacific Region* (ed. D.G. Howell), Circum-Pacific Council for Energy and Mineral Resources, Earth Sciences Series, vol. 1, AAPG Bookstore, Tulsa, Oklahoma, pp. 515–521.

Bourgois, J., Pautot, G., Bandy, W. *et al.* (1988) Seabeam and seismic reflection imaging of the tectonic regime of the Andean continental margin off Peru (4° S to 10° S). *Earth Planet. Sci. Lett.* **87**, 111–126.

Bourgois, J., Toussaint, J.-F., Gonzalez, H., *et al.* (1987) Geological history of the Cretaceous ophiolitic complexes of northwestern South America (Colombian Andes). *Tectonophysics* **143**, 307–327.

Burton, R., Kendell, C.G.S., and Lerche, I. (1987) Out of our depth; on the impossibility of fathoming eustasy from the stratigraphic record. *Earth Sci. Rev.* **24**, 237–277.

Cadet, J.P. *et al.* (1985) Oceanographie dynamique. *C. R. Acad. Sci. Paris Ser. 2*, **301**, 287–296.

Cadet, J.P. *et al.* (1987) Deep scientific dives in the Japan and Kuril Trenches. *Earth Planet. Sci. Lett.* **83**, 313–328.

Card, K.D. (1986) Tectonic setting and evolution of Late Archean greenstone belts of Superior province, Canada, in *Tectonic Evolution of Greenstone Belts* (eds M. J. deWit and L.D. Ashwal). Lunar and Planetary Sciences Institute Tech. Rep. 86–10, Houston, pp. 74–76.

Champion, D.E., Howell, D.G., and Grommé, C.S. (1984) Paleomagnetic and geologic data indicating 2500 km of northward displacement for the Salinian and related terranes, California. *J. Geophys. Res.* **89**, 7736–7752.

Champion, D.E., Howell, D.G., and Marshall, M. (1986) Paleomagnetism of Cretaceous and Eocene strata, San Miguel Island, California Borderland, and the northward translation of Baja California. *J. Geophys. Res.* **91**, 11 557–11 570.

Chase, C. (1978) Plate kinematics: the Americas, east Africa, and the rest of the world. *Earth Planet. Sci. Lett.* **37**, 355–368.

Choukroune, P., Lopez-Munoz, M., and Quale, J. (1983) Cisaillement ductile sud-americain et déformations discontinues associées: mise en évidence de la déformation régionale non coaxiale dextre. *C. R. Acad. Sci. Paris* **296**, 657–660.

Coe, R.S., Globerman, B.R., Plumley, P.W., and Thrupp, G.A. (1985) Paleomagnetic results from Alaska and their tectonic implications, in *Tectonostratigraphic Terranes of the Circum-Pacific Region* (ed. D.G. Howell) Circum-Pacific Council for Energy and Mineral Resources, Earth Science Series, vol. 1, AAPG Bookstore, Tulsa, Oklahoma pp. 85–120.

Condie, K.C. (1984) *Plate Tectonics and Crustal Evolution*, 2nd edn, Pergamon Press, Oxford.

Coney, P.J. (1981) Accretionary tectonics in Western North America, *Arizona Geol. Soc. Bull.* **15**, 23–37.

Coney, P.J. and Jones, D.L. (1985) Accretion tectonics and crustal structure in Alaska. *Tectonophysics* **119**, 265–283.

Coney, P.J., Jones, D.L., and Monger, J.W.H. (1980) Cordilleran suspect terranes. *Nature* **239**, 329–333.

Cox, A. (1973) *Plate Tectonics and Geomagmetic Reversals*, W.H. Freeman and Company, San Francisco.

Cox, A. and Hart, R.B. (1987) *Plate Tectonics; How it Works*, Blackwell Scientific Publications, Oxford.

Davis, D., Suppe, J., and Dahlen, F.A. (1983) Mechanics of fold-and-thrust belts and accretionary wedges. *J. Geophys. Res.* **88**, 1153–1172.

DePaolo, D.J. (1984) The mean life of continents; estimates of continental recycling from Nd and Hf isotopic data and implications for mantle structure. *Geophys. Res. Lett.* **10**, 705–708.

Dercourt, J. Zoenshain, L.P., and Ricon, L.E. (1986) Geological evolution of the Tethys belt from the Atlantic to the Pamirs since the Lias. *Tectonophysics* **123**, 241–315.

Engebretson, D.C. (1982) Relative motion between oceanic and continental plates in the Pacific Basin. PhD dissertation, Stanford University, California, USA.

Engebretson, D.C., Cox, A., and Gordon, R.G. (1984) Relative motion between oceanic plates of the Pacific Basin. *J. Geophys. Res.* **89**, 10 291–10 310.

Ford, M. (1987) Practical application of the sequential balancing technique: an example from the Irish Variscides. *J. Geol. Soc. London* **144**, 885–891.

Freund, R. (1974) Kinematics of transform and transcurrent faults. *Tectonophysics* **21**, 93–134.

Funk, H., Oberansli, R., Pfiffner, P. *et al.* (1987) The evolution of the northern margin of Tethys in eastern Switzerland. *Episodes* **10**, 102–107.

Fyfe, W.S. (1978) The evolution of the Earth's crust; modern plate tectonic to ancient hot spot tectonic? *Chem. Geol.* **23**, 89–114.

Glen, W. (1982) *The Road to Jaramillo*, Stanford University Press, Stanford, California, USA.

Gordon, R.G. (1988) True polar wander and paleomagnetic poles. *Phys. Today* **41**, S-46.

Groves, D.I., Ho, S.E., Rock, N.M.S. *et al.* (1987) Archean cratons, diamonds and platinum; evidence for coupled long-lived crust–mantle systems. *Geology* **15**, 801–805.

Hamilton, J. (1979) Tectonics of the Indonesian region. *US Geol. Surv. Profess. Paper* 1078.

Helwig, J. (1974) Eugeosynclinal basement and a collage concept of orogenic belts, in *Modern and Ancient Geosynclinal Sedimentation* (eds R.H. Dott and R.H. Shaver), *Soc. Econ. Paleontol. Mineral. Spec. Publ.* **19**, 359–376.

Hess, H.H. (1962) History of ocean basins, in *Petrologic Studies – a Volume to Honor A.F. Buddington* (eds A.E.J. Engel *et al.*), Geological Society of America, Boulder, Col., pp. 599–620.

Hildebrand, R.S., Hoffman, P.F., and Bowing, S.A. (1987) Tectono-magmatic evolution of the 1.9 Ga Great Bear magmatic zone, Wopmay orogen, northwestern Canada. *J. Volcanol. Geotherm. Res.* **32**, 99–118.

Hillhouse, J.W. (1977) Paleomagnetism of the Triassic Nikolai Greenstone, McCarthy Quadrangle, Alaska. *Canad. J. Earth Sci.* **14**, 2578–2592.

Hillhouse, J.W. and Grommé, C.S. (1984) Northward displacement and accretion of Wrangellia; new paleomagnetic evidence from Alaska. *J. Geophys. Res.* **89**, 4461–4477.

Hoffman, P.F. (1987) Continental transform tectonics: Great Slave Lake shear zone (ca. 1.9 Ga), northwest Canada. *Geology* **15**, 785–788.

Hoffman, P.F. and Bowing, S.A. (1984) Short lived 1.9 Ga continental margin and its destruction Wopmay orogen, northwest Canada. *Geology* **12**, 68–72.

Hoffman, P.F., Dewey, J.F., and Burke, K. (1974) Aulacogens and their genetic relation to geosynclines, with a Proterozoic example from Great Slave Lake, Canada, in *Modern and Ancient Geosynclinal Sedimentation* (eds R.H. Dott and R.H. Shaver). *Soc. Econ. Paleontol. Mineral. Spec. Publ.* **19**, 38–55.

Hornafius, J.S. (1985) Neogene tectonic rotation of the Santa Ynez Range, Transverse Ranges, California, suggested by paleomagnetic investig-ation of the Monterey Formation. *J. Geophys. Res.* **90**, 12 503–12 522.

Howell, D.G. (1980) Mesozoic accretion of exotic terranes along the New Zealand segment of Gondwanaland. *Geology* **8**, 487–491.

Howell, D.G., Gibson, J.D., Fuis, G.S. *et al.* (1985) Pacific abyssal plain to the Rio Grande rift. *Geol. Soc. Amer. Centenn. Continent/Ocean Transect* **5**, C-3.

Howell, D.G. and Murray, R.W. (1986) A budget for continental growth and denudation. *Science* **233**, 446–449.

Howell, D.G. and Wiley, T.J. (1987) Crustal evolution of northern Alaska inferred from sedimentology and structural relations of the Kandik basin. *Tectonics* **6**, 619–631.

Hsü, K.J. (1981) Thin-skinned plate-tectonic model for collision-type orogensis. *Sci. Sin.* **24**, 100–110.

Hurley, P.M. and Rand, J.R. (1969) Pre-drift continental nuclei. *Science* **164**, 1229–1242.

Irving, E., Monger, J.W.H., and Yole, R.W. (1980) New paleomagnetic evidence for displaced terranes in British Columbia, in *The Continental*

Crust and its Mineral Deposits (ed. D.W. Strangway). *Geol. Assoc. Canada Spec. Pap.* **20**, 441–456.

Ji, X. and Coney, P.J. (1986) Accreted Terranes of China, in *Tectonostratigraphic terranes of the Circum-Pacific Region* (ed. D. G. Howell). Circum-Pacific Council for Energy and Mineral Resources, Earth Science Series, vol. 1, AAPG Bookstore, Tulsa, Oklahoma, pp. 349–361.

Jolivet, L., Huchon, P., and Rangin, C. (1989) Tectonic setting of western Pacific marginal basins. *Tectonophysics*, in press.

Jones, D.L., Silberling, N.J., and Coney, P.J. (1986) Collision tectonics in the Cordillera of western North America, in *Collision Tectonics* (eds M.P. Coward and R.C. Alison), Blackwell Scientific Publications, Oxford, pp. 367–387.

Jones, D.L., Silberling, N.J., Coney, P.J., and Plafker, G. (1987) Lithotectonic terrane map of Alaska. *US Geol. Surv.* Map MF-1874-A, scale 1:2 500 000.

Kanter, L.R. and Debiche, M. (1985) Modeling the motion histories of the Point Arena and central Salinia terranes, in *Tectonostratigraphic Terranes of the Circum-Pacific Region* (ed. D.G. Howell). Circum–Pacific Council for Energy and Mineral Resources, Earth Science Series, vol. 1, AAPG Bookstore, Tulsa, Oklahoma, pp. 227–238.

Kay, M. (1951) North American geosynclines. *Geol. Soc. Amer. Mem.* **48**, 1–148.

King, P. (1966) Tectonic map of North America. *US Geol. Surv.* scale 1:5 000 000.

Kipp, M.E. and McLosh, H.J. (1986) A preliminary numerical study of colliding planets, in *Origin of the Moon* (eds W.K. Hartmen *et al.*), pp. 643–647, Lunar and Planetary Sciences Institute, Houston.

Kuenen, P.H. (1939) Quantitative estimates relating to eustatic movements. *Geologie en Mijnbouw* **8**, 194–201.

Lallemand, S. (1987) La Fosse du Japon. PhD Dissertation, Université d'Orleans, Orléans, France.

Lallemand, S., Cullota, R., and von Huene, R. (1989) Subduction of the Kashima Seamount in the Japan Trench. *Tectonophysics*, in press.

Lee, T.Q. (1983) Focal mechanism solutions and their tectonic implications in Taiwan region. *Acad. Sin. Inst. Earth Sci. Bull.* **3**, 37–54.

Luyendyk, B.P., Kammerling, M.J., Terres, R.R., and Hornafius, J.S. (1985) Simple shear of southern California during Neogene time suggested by paleomagnetic declinations. *J. Geophys. Res.* **90**, 12 454–12 466.

Lyberis, N. (1985) Tectonic evolution of the north Aegean Trough, in *The Geological Evolutions of the Eastern Mediterranean* (eds J.E. Dixon and A.H.F. Robertson), *Geol. Soc. London Spec. Publ.* **17**, 709–725.

McCarthy, J. and Scholl, D.W. (1985) Mechanisms of subduction accretion along central Aleutian Trench. *Geol. Soc. Amer. Bull.* **96**, 691–701.

McElhinny, M.W. (1973) *Paleomagnetism and Plate Tectonics*, Cambridge University Press, Cambridge, UK.

McKenzie, D.P. (1970) Plate tectonics of the Mediterranean region. *Nature* **226**, 239–243.

McLennan, S.M. and Taylor, S.R. (1982) Geochemical constraints on the growth of the continental crust. *J. Geol.* **90**, 347–361.

Meissner, R., Wever, T., and Fluh, E.R. (1987) The Moho of Europe – implications for crustal development. *Ann. Geophys.* **5**, 357–364.

Minster, J.B. and Jordan, T.H. (1978) Present-day plate motions. *J. Geophys. Res.* **83**, 5331–5354.

Monger, J.W.H., Clowes, R.M., Price, R.A. *et al.* (1985) Juan de Fuca plate to Alberta plains. *Geol. Soc. Amer. Centenn. Continent/Ocean Transect* **7**, B-2.

Nelson, B.K. and DePaolo, D.J. (1985) Rapid production of continental crust 1.7–1.9 G.y. ago; Nd and Sr isotopic evidence from the basement of the North American midcontinent. *Geo. Soc. Amer. Bull.* **96**, 746–754.

Oxburgh, E.R. (1972) Flake tectonics and continental collision. *Nature* **239**, 202–204.

Ozawa, T. and Kanmera, K. (1984) Tectonic terranes of late Paleozoic rocks and their accretionary history in the Circum-Pacific Region viewed from fusulinacean paleobiogeography. *Stanford Univ. Publ. Geol. Sci.* **18**, 158–160.

Pachett, J. and Chauvel, C. (1984) The mean life of continents is currently not constrained by Nd and Hf isotopes. *Geophys. Res. Lett.* **11**, 151–153.

Pachett, P.J. and Arndt, N.T. (1986) Nd isotopes and tectonics of 1.9–1.7 Ga crustal genesis. *Earth Planet. Sci. Lett.* **78**, 329–338.

Pachett, P.J., White, W.M., Feldman, H. *et al.* (1984) Hafnium–rare earth element fractionation in the sedimentary system and crustal recycling into the Earth's mantle. *Earth Planet. Sci. Lett.* **69**, 365–378.

Panuska, B.C. and Stone, D.B. (1985) Latitudinal motion of the Wrangellia and Alexander terranes and the southern Alaska superterrane, in *Tectonostratigraphic Terranes of the Circum-Pacific Region* (ed. D.G. Howell). Circum-Pacific Council for Energy and Mineral Resources, Earth Science Series, vol. 1, AAPG Bookstore, Tulsa, Oklahoma, pp. 109–120.

Pelletier, B. and Stephan, J.F. (1986) Middle Miocene obduction and late Miocene beginning of collision registered in the Hengchun Peninsula; geodynamic implications for the evolution of Taiwan. *Geol. Soc. China Mem.* **7**, 301–324.

Price, N.J. and Audley-Charles, M.G. (1987) Tectonic collision processes after plate rupture: *Tectonophysics* **140**, 121–129.

Reymer, A. and Schubert, G. (1984) Phanerozic addition rates to the continental crust growth. *Tectonics* **3**, 63–77.

Roeder, D., Gilbert, O.E. Jr, and Witherspoon, W.D. (1978) Evolution and macroscopic structure of Valley and Ridge thrust belt, Tennessee and Virginia. *Univ. Tennessee Dept. Geol. Studies in Geol.* **2**, 1–25.

Roure, F., Choukroune, P. *et al.* (1989) ECORS deep seismic data and balanced cross-sections: geometric constraints to trace the evolution of the Pyrenees, *Tectonics* **8**(1), 41–50.

Scharer, U., Krough, T.E., and Gower, C.F. (1986) Age and evolution of the Grenville province in eastern Labrador from U–Pb systematics in accessory minerals. *Contrib. Mineral. Petrol.* **94**, 438–451.

Scholl, D.W. and Ryan, H.F. (1986) Crustal structure and evolution of the Aleutian forearc in the vicinity of the Andreanof earthquake. *Eos,* **67**, 1081.

Scholl, D.W., Vallier, T.L., and Stevenson, A.J. (1987) Geologic evolution and petroleum potential of the Aleutian Ridge, in *Geology and Resource Potential of the Continental Margin of Western North America and Adjacent Ocean Basins – Beaufort Sea to Baja California* (eds D.W. Scholl *et al.*). Circum-Pacific Council for Energy and Mineral Resources, Earth Science Series, vol. 6, AAPG Bookstore, Tulsa, Oklahoma, pp. 123–155.

Schuchert, C. (1916) Correlation and chronology on the basis of paleogeography. *Geol. Soc. Amer. Bull.* **27**, 419–513.

Schuchert, C. (1923) Sites and natures of the North American geosynclines. *Geol. Soc. Amer. Bull.* **34**, 151–260.

Sclater, J.C., Parsons, S., and Jaupart, C. (1981) Oceans and continents; similarities and differences in the mechanisms of heat loss. *J. Geophys. Res.* **86**, 11 535–11 552.

Scotese, C.R. (1984) Paleozoic paleomagnetism and the assembly of Pangea, in *Plate Reconstructions from Paleozoic Paleomagnetism* (eds R. Van der Voo *et al.*). Amer. Geophys. Union Geodyn. Ser. vol. 12, pp. 1–10.

Scotese, C.R., Bambach, R.K., Barton, C. *et al.* (1979) Paleozoic base maps: *J. Geol.* **87**, 217–277.

Sengor, A.M.C. (1984) The Cimmeride orogenic system and the tectonics of Eurasia. *Geol. Soc. Amer. Spec. Publ.* **195**, 1–82.

Sengor, A.M.C. (1987) Tectonics of the Tethysides; orogenic collage development in a collisional setting. *Ann. Rev. Earth Planet. Sci.* **15**, 213–244.

Shipley, T.H. and Moore, G.F. (1986) Sediment accretion, subduction, and dewatering at the base of the trench slope off Costa Rica: a seismic reflection view of the decollement. *J. Geophys. Res.* **91**, 2019–2028.

Silberling, N.J. (1985) Biogeographic significance of the Upper Triassic bivalve *Monotis* in Circum-Pacific accreted terranes, in *Tectonostratigraphic Terranes of the Circum-Pacific Region* (ed. D.G. Howell). Circum-Pacific Council for Energy and Mineral Resources, Earth Science Series, vol. 1, AAPG Bookstore, Tulsa, Oklahoma, pp. 63–70.

Silberling, N.J., Jones, D.L., Blake, M.C. Jr, and Howell, D.J. (1987) Lithotectonic terrane map of the western conterminous United States: *US Geol. Surv. Map* MF-1874-C, scale 1:2 500 000.

Speed, R.C. and Westbrook, G.K. (1984) Lesser Antilles arc and adjacent terranes. *Marine Sci. Int. Ocean Margin Drilling Program, Regional Atlas* **10**, 27 sheets.

Stauffer, P.H. (1985) Continental terranes in southeast Asia: pieces of which puzzle? in *Tectonostratigraphic Terranes of the Circum-Pacific Region* (ed. D.G. Howell). Circum-Pacific Council for Energy and Mineral Resources, Earth Science Series, vol. 1, AAPG Bookstore, Tulsa, Oklahoma, pp. 529–539.

Suppe, J. (1980) A retrodeformable cross section of northern Taiwan. *Geol. Soc. China Proc.* **23**, 46–55.

Taira, A., Tokuyama, H., and Soh, W. (1989) Accretion tectonics and evolution of Japan, in *The Evolution of the Pacific Ocean Margins* (ed. Z. Ben-Avraham), Oxford University Press, Oxford.

Tapponnier, P., Mattauer, M., Proust, F., and Cassaigneau, C. (1981) Mesozoic ophiolites, sutures, and large-scale tectonic movements in Afghanistan. *Earth Planet. Sci. Lett.* **52**, 355–371.

Tapponnier, P., Peltzer, G., and Armijo, R. (1986) On the mechanics of the collision between India and Asia, in *Collision tectonics* (eds M.P. Coward and R.C. Alison), Blackwell Scientific Publications Oxford, pp. 115–158.

Tarduno, J.A., McWilliams, M., Sliter, W.V. *et al.* (1986) Southern hemisphere origin of the Cretaceous Laytonville Limestone of California. *Science* **231**, 1425–1428.

Tarling, D.H. (1981) *Paleomagnetism*, Chapman and Hall, London.

Taylor, S.R. and McLennan, S.M. (1985) *The Continental Crust; its Composition and Evolution*, Blackwell Scientific Publications, Oxford.

Uyeda, S. (1987) *The New View of the Earth*, W.H. Freeman, San Francisco.

Veizer, J. and Jansen, S.L. (1979) Basement and sedimentary recycling and continental evolution. *J. Geol.* **87**, 341–370.

Veizer, J. and Jansen, S.L. (1984) Basement and sedimentary recycling; time dimension to global tectonics. *J. Geol.* **93**, 625–643.

Viallon, C., Huchon, P., and Barrier, E. (1986) Opening of the Okinawa basin and collision in Taiwan: a retreating trench model with lateral anchoring. *Earth Planet. Sci. Lett.* **80**, 145–155.

Vine, F.J. and Mathews, D.H. (1963) Magnetic anomalies over oceanic ridges. *Nature* **199**, 947–949.

Vink, G.E., Morgan, W.J., and Wu-Ling, Z. (1984) Preferential rifting of continents: A source of displaced terranes. *J. Geophy. Research* **89**, B12, 10 072.

Wegener, A. (1912) Die Entstehung der Kontinente. *Geol. Rundschau* **3**, 276–292.

Wegener, A. (1924) *The Origin of Continents and Oceans*, Methuen, London.

Williams, H. *et al.* (1989) Anatomy of North America; Thematic Geologic Portrayals of the Continent. *Tectonophysics.* Special volume commemorating 25th anniversary of plate tectonics, tectonophysics, Amsterdam.

Wilson, J.T. (1965) A new class of faults and their bearing on continental drift. *Nature* **207**, 343–347.

Wise, D.U. (1974) Continental margins, freeboard and the volumes of continents and oceans through time, in *The Geology of Continental Margins* (eds C.A. Burk and C.L. Drake), Springer-Verlag, New York, pp. 45–48.

Wood, J.A. (1986) Moon over Mauna Loa, a review of hypotheses of formation of Earth's moon, in *Origin of the Moon* (eds W.K. Hartmen *et al.*), Lunar and Planetary Sciences Institute, Houston, pp. 17–55.

Xu, J., Zhu, G., Tong, W., Cui, K., and Liu, Q. (1987) Formation and evolution of the Tancheng–Lujiang wrench fault system: a major shear system to the northwest of the Pacific Ocean. *Tectonophysics* **134**, 273–310.

Zhu, Z. and Teng, J. (1984) Preues paleomagnetiques de la derive vers le Nord de fragments de la plaque indienne apres la separation du continent de Gondwana et de leur collision avec la plaque, in *Mission Franco–Chinoise a Tibet* (eds J.L. Mercier and Li Guangcen), CNRS, Paris, France, pp. 15–20.

Zonenshain, L.P., Kononov, M.V., and Savostin, L.A. (1989) Pacific and Kula/Eurasia relative motions during the last 130 Ma and their bearing on orogenesis in northeast Asia. *Amer. Geophys. Union Geodyn. Ser.* **18**, 29–47.

Index

Abyssal plains 32
Accretion 89, 105
  prism 37
  tectonics 35, 80
  efficiency 76
  wedge 39
Afghanistan 170
Africa 112
Agulhas 12, 13
Alaska 51, 181
  mount McKinley (Denali) 186
Alaskan shelf 75
Aleutian 73
  accretionary prisms 76
  arcs 73
  trench 26, 75
Aleutian-Komandorsky islands 42
Alexander terrane 79
Allochthonous 121
  crustal fragments 81
Alpine fault 16, 151
Alps 15, 29, 157, 172
Altiplano 16
Amalgamate 89, 105, 162
Amitsoq Orthogneiss 5
Amlia islands 73
Anadyr-Koryak mountains 42
Anatectic granite 186
Anatolian 16, 152
Andaman islands 27
Andaman sea 117
Andean magmatic arc 117
Andes 16, 17, 29, 40, 157, 195
Anglessy melange 89
Angular momentum 59
Antarctica 12
Apennines 29

Appalachian 119
Appalachian-Caledonian foldbelt 45
Appalachian-Caledonian system 44
Apparent polar wander (APW) 46, 50,
    51
Archean 3, 25, 57, 61, 95, 111
Archipelagoes 50
Arcs
  mariana 41
  island 34
  volcanic 34
  tonga 41
Arctic islands 199
Arc-continent collisions 35
Arequipa massif 197
Arizona 194
Asthenosphere 28
Athapuscow aulocogen 115
Aulocogens 115
Australia 112
Autochthonous 121
Avalon terrane 88
Aves, see Plateaus
A-subduction 29, 40, 157, 163, 172,
    181, 186, 191, 198, 199

Baberton Mountain Land 112
Backarc basins 41
Baja California 138
Baja-BC arc 188
Baja-borderland composite terrane 189,
    191
Banda Sea 157, 167
Barbados 69
Barracuda Ridge 73
Basalt 7
Basin and Range province 12, 13, 189

Bay of Biscay 143
Belt 86
Bengal Sea 172
Benue trough 115
Bering sea 41
Blind thrust 15
Block 84
Borneo 157
Bowers, *see* Plateaus
British Columbia 186
Broken ridge, *see* Plateaus
Brooks range 15, 79, 143, 153, 182, 183
Buoyancy forces 40
Burma 77
B-subduction 23, 30, 35, 41, 164, 186, 188

Caledonian, *see* Organic belts
California 9
    Coast Ranges 9, 93
    continental borderland 189
Canada 32
    Rockies 153
Canary islands, *see* Plateaus
Canyon Diablo meteorite 3
Caples 151
Caribbean
    plate 19
    sea 16
Carnegie, *see* Plateaus
Central belt 89
Central belt terrane 142
Chile 28, 32
Chilliwack batholith 97, 99
Chromite 84
Chuckanut formation 97, 99
Chugach terrane 88
Churchill province 20, 117
Cimmeria 45, 93
Cimmerian continent 93
Circum-pacific 77
Coast plutonic complex 188
Coast range thrust 201
Colombia 196
Colville plain 182
Composite terrane 97
Compressional mountain systems 13

Conjugate margins 13
Continental
    crust recycling 54
    denudation 76
    drift 51
    freeboard viii, 57
    growth viii, 19, 35, 80
    lithosphere 167
    slivers 164
        quartz 167
        olivine 167
Continents 2
    composition 2
    Eurasia 169
Convection cells 28
Cordillera, *see* Organic belts
Coronation super group 117
Critical taper 37
Crust
    continental 1
    oceanic 1
    Earth's 2

Darwin, Charles 145
Dating 204
    U 204
    Pb 204
    K 204
    Ar 204
    Sm 204
    Nb 204
    Rb 204
    Sr 204
Decollement 13, 37
Deep sea
    clay 71
    drilling program (DSDP) 2, 71, 200
    mud 3
Denali 152
Detachment faults 155
Dewatering 37
Dietz, Robert 22
Dikes 7
Dispersion tectonics 16, 89, 106, 107, 162
Distributed shear 152
Drag forces 28

Dun Mountain 88
Dunnage terrane 88
Duplex 100
Dynamic 7

Earth's magnetic field 45
  blocking temperature 46
  thermal remanent magnetization 46
  chemical remanent magnetization 46
  detrital chemical remanent
    magnetization 46
  magnetic declination 46
  magnetic inclinations 46
  magnetostratigraphy 46
Earth's spin axis 48
Earth's surface 9
East African rift system 18, 115
East Pacific Rise 13
Ebro basin, *see* Pyrenees
Echo-sounding systems 22
Eclogite 66
Elevation 9
Euler pole 26, 49, 125, 131
Eurasia 27, 35, 73, 110, 169
European alps 111, 121, *see* Organic
  belts
Exmouth, *see* Plateaus

Fairweather-Denali fault 90
Fairweather-Queen Charlotte transform
  fault 74
Fixist 81
Flakes
  obducted sheets 84
Fluid pressure 37, 38
Flysch basins 80
Fold-and-thrust belt 40
Forearc basin 37
Fort Simpson volcanic arc 117
Franciscan 101, 110, 142, 155, 201
Freeboard 60, 63
Fusulinids
  Schwagerinidae 146
  Verbeekinidae 146

Gabbro 66
Gabbroic magma chambers 7

Galapagos Rise 34
Genetic 95
Geochemistry 67
Geodetic survey 124
Geohistory 100
Geometric analysis 5
Geothermal gradient 66
Global budget 76
Global-circuit analysis 129
Glossopteris 146
Golconda allochthon 153
Gondwana (land) 45, 49, 50, 91, 100,
  119, 138, 140, 146, 177, 195, 197
Gorgona Island 197
Gravitational body forces 28
Gravitational pull 9
Great basin of Western North
  America 155
Great Bear magmatic arc 117
Great Slave lake 117
Great Valley sequence 201
Greenland 168
Greenstone belts 86, 111, 112
Grenville Province 93, 119
  orogeny 95
  basement 95
Growth rates 34
Guadalcanal Island 28
Gulf of Aden 115
Gulf of California 151
Gulf of Mexico 194, 199

Haast Schist 99
Harry Hess 22
Hawaiian-Emperor chain 17
  see also Plateaus
Hawaiian-Emperor seamount 168
Hepburn intrusives 117
Hercynian
  orogeny 173
  sutures 173
Hess rise 34
Hikurangi trough 151
Himalaya 29, 157
Himalayan foldbelt 167
Hokkaido 71
Hokonue 88, 151

Hotspots 17, 65, 122, 131
  Canary 7
  Iceland 17
  Hawaiian 17
  Mount Kilimanjaro 17
  Tahitian Islands 17
Hottah 119

Iapetus 45
Iberian Peninsula 77, 155
Impacting planetesimals 59
India 104, 138
  Ganges plain 171, 172
Indo-Australia 27
Innuition, *see* Organic belts
Internal friction 37
Ion probe 5
Island arcs 34
Isotherm 14000 25, 29
Isotopes 3
  ^{144}Nd 65
  ^{176}Lu 68
  ^{177}Hf 68
  ^{207}Pb/^{204}Pb 68
  ^{87}Sr/^{86}Sr 99, 198, 204
  Beryllium-10 68
  ε_{Nd} 65
  K/Ar 54
  Rb/Sr 54
  Sm/Nd 20
  U/Pb 54
Isua supercrustal sequence 5

Japan 35, 41, 51
Jura mountains 15

Kaltag fault 152
Kerguelen 12, 101
Kinematic analysis 5
  linkages 95
Knocker 84, 101
Komandorsky Islands 26
Komatiites 60
Komatiite lava flows 197
Kula, *see* Plates
Kurosegawa terrane 111

Luzon 43
Labrador 93
Labrador sea, *see* Seas
Laramide orogeny 188, 191
Late Jurassic 73
Lau basin 41
Laurasia 93
Lesser Autilles arc system 37
  accretionary prism 73
Light rare earth patterns 64
Lord Howe rise 12
  see also Plateaus
Louisville ridge 34
  see also Plateaus
Lunar Highlands 5
Lungless salamanders 145

Magnetic
  lineations 23, 124
  anomalies 23
  reverse or normal 23
  pole 50
Makkovik province 93
Manihiki, *see* Plateaus
Manila trench 164
Mariana 71, 76
Massifs 15
Mazatzal 117
Mendocino transform 151
Metamorphic isograds 132
Meteorites 3
Microplates 50, 82
Midocean ridges 7, 23, 24, 29, 31, 32, 60
Mid-Atlantic ridge system 18, 191
Mid-Pacific mountains 34
Mindoro Islands 164
Mino terrane 153
Moho drilling project 1
Mojave-Sonora Megashear 194
Mollweide transverse oblique map 4
Monotis 148
Moon 3, 5
Mountain building 80'
Mountains 8

Nanaimo group 99

Nankai trough 69
Nappe 82
  double fold of Glarus 82
  Helvetic 178
  Penninic 178
  Austral Alpine 178
Nazca, *see* Plateaus
New Guinea 146, 199
New Hebrides 28, 34
New Zealand 9, 26, 93, 139
  Mount Cook 9
  southern 9
New-Tethys Sea 45
Ninety East ridges, *see* Plateaus
North American plate 126
North Sea 199
Northern Mariana 35

Ocean drilling project (ODP) 2, 13, 15,
    67, 69, 71
Ocean floor basalt 23
Oceanic crust recycling 52
Oceanic lithosphere 60
Oceanic plateaus, *see* Plateaus
Oceanic sediment 54
Oceanic terranes 80
Oceans 6
Ocean-continent boundary 167
Ogaswara plateau, *see* Plateaus
Okinawa trough 43
Omolon 77
Ontong Java plateau, *see* Plateaus
Ophiolitic 32, 112
Organic belts 88
  Caledonian/Appalachian
  calidonides 88
  Cordillera 29, 91, 93, 105, 107, 119,
      127, 130, 137, 157, 181, 194
  European alps 93
  Innuition 119
  Ouachitan 119
  Tethyan 91, 110, 130
  Tethysides 88
Orogeny 45, 117
Oronoco river 71
Ouachitan, *see* Organic belts
Overlap sequence 97

Pacific-Kula spreading ridge lay 73
Palau-Kyushu ridge, *see* Plateaus
Palau-kyushu ridge 43
Paleo 45
Paleogeographic reconstructions 80
Paleomagnetic data 45, 50
Paleomagnetism 132
  cryogenic magnetometers 132
Pangea viii, 26, 43, 79, 106, 112, 173,
    181
Parautochthonous 121
Peruvian 16
Phanerozoic 5, 57, 61
Philippines 77, 157
Piggyback basins 164
Plate tectonic reconstructions 43
Plate tectonics 80
Plateaus 33
  Agulhus 12, 101
  Aves 102
  Bowers 102
  Broken ridge 101
  Canary Islands 101
  Carnegie 104
  Exmouth 12, 101
  Hawaiian-Emperor 101
  Kerguelen 12, 101
  Lomonsof ridge 101
  Lord Howe Rise 12, 101
  Louisville 101
  Louisville ridges 104
  Masquerine 33
  Nazca 104
  Ninetyeast ridges 101
  Oceanic Plateaus
    Manihiki 102
    Ontong Java 19, 33, 102, 104
  Ogasawara 104
  Palau-Kyushu ridges 102
  Shirshov ridge 104, 172
  Umnak Plateau 104, 172
  Walrus 101
Plates, the nature of 25
  Africa 26, 126, 179
  Antarctica 26, 126
  Arabia 26
  Caribbean 26

Plates *contd*
　Cocos 26
　Eurasia 26, 27, 44, 50, 124, 169
　European 179
　Farallon 126
　Indonesian 50
　Indo-Australia 26, 83
　Kula 130, 143
　Nasca 26
　North America 26, 50, 124, 126
　Pacific 26, 126
　Philippine 26, 50, 160
　Scotia 26
　South America 26
Plethodontidae 145
Porcupine river 181
Precambrian terranes 111
Present day seismicity 124
Principal driving forces 28
　viscous-drag 26
Proterozoic 5, 18, 111, 112
Proto San Andreas 151
Provinancial linkage 99
Province 86
Prudoe Bay 199
Pull-apart basins 41
Pure shear 13
Puysegur trench 151
Pyrenees 16, 121, 157, 173
　North Pyreneean fault system
　　zone 173
　Aquitane basin 175
　Ebro basin 175, 173

Queen Charlotte fault system 189
Queen Maud uplift 117

Rate of denudation 76
Recycling rate 54
Red Sea 115
Reykjanes 168
Ridge push 29, 172, 191
Rifts 89
Rio Grande rift 13, 189, 194
Rocky mountains 12
Russia 79

Subducting continental crust 66

Subduction 40
　erosion 40
　zones 24, 28, 34, 35, 39, 63, 71
Submarine-fan 88
Sulu arc 164
Sundaland 77, 87
Suspect 81
Suspect terrane 105, 186, 202
Sutures 115
System 27
　orogenic 27
　Caledonide 27
　Tyrrhenian 179
Seismic stratigraphy 69
Seismic-reflection data 66
Seward Peninsula 79
Shimanto terrane 88
Shirshov ridge, *see* Plateaus
Sialic material 35
Sierra Nevada 12
Sills 7
Simple shear 13
Size of terranes 100
Slab pull 29, 172
Slave 119
　craton 117
　province 112, 115
Silvers 85
Snake river plain 17
Solar-system accretion 59
Solomon arc 104
Sonora 194
South Keewatin 119
Southern Alps 93
Southern California boderland 16
Sri Lanka 29
State-of-stress 203
Stitching plutons 97
Straight Creek fault 97, 189
Stratigraphic terranes, *see* Terranes
Strike-slip faulting 71
Subcretion 37, 71
Sala Y Gomex ridge 34
Salton trough 189
San Andreas fault 87, 90, 151
　system 191, 192
San Juan Islands 99

Santa Lucia-Orocopia composite
    terrane 189, 191
Satellite Laser Ranging 122
Scandinavia 28
Scotland 27
Scripps Institution of Oceanography 23
Seamounts 82
Seas 41
    Banda 43
    Bering 42
    Black 43
    Caribbean 42
    Caspian 42
    Celebes 43
    Japan 41
    Labrador 168
    Neo-Tethys 45
    Scotia 42
    South China 41, 163
        Reed Bank 164
        Dangerous Ground 164
    Sulu 43

Taiwan 35, 153, 157
    Taiwan coastal range 161
Tan Lu fault 152
Tectonic Assemblage diagram 97
Tectonic erosion 197
Tectonostratigraphic vii
Tectonostratigraphic terranes viii, 6, 21,
    82, 86
Tensional mountain systems 9
Terrain viii, 87
Terrane viii, 87
    Angayucham 184
    archean 112
    Birch Creek Schist 89
    boundaries 100
    distributed 88
    endicott 184
    Hammond 184
    map 104
    metamorphic 89
    stratigraphic 87
    superior province 115, 119
Tethyan foldbelt 121
Tethyan Sea 121

ocean 146
Tethyan 110
Tethys 173
Thelon fault 117
Thelon magmatic zone 117
Thermal anomalies 9
Thermal welts 17
Thermally thickened crust 18
Thick-skinned 15, 16
    thrusting 41, 153
    duplexing 153
Thrust faults 153
Tibet 88, 117, 138, 167
Tibetan plateau 15, 157, 170
Tiburon rise 71, 73
Timor 157, 161
Tintina 90, 189
Torlesse terrane 88, 99
Transantarctic mountains 12
Transcurrent faults 148
    piercing points 148
Transcurrent mountain systems 16
Transform faults 24
Transhimalaya range 88
Transpression 16, 30, 191
Transtension 16, 30, 191
Transverse ranges of 143
Trans-Labrador batholith 93
Trapped oceanic crust 42
Trench 37, 40
    Peru-Chile 40
    Middle America 40
    rollback 43
    suction 43
Trench-slope basins 164
Turkey 51, 77
Tuzo Wilson, J. 24, 44

Ultramafic rock 7
Umnak plateaus, *see* Plateaus
Underplating 37, 71

Valley of Ten Thousand Smokes 17
Vancouver Island 32, 99
Very long baseline interfacing 122
Volatiles 59
Volcanoes 180

Volcanoes *contd*
  Mt Etna 180
  Mt Vesuvius 180

Wadati-Benioff zones 23, 35, 41, 191
Wallace, Alfred Russell 145
Walrus, *see* Plateaus
Wegener, A. 24
Welding metamorphism 99
West Antarctica 79
Whole-rock age 54

Wilson cycle viii, 44, 48
Wind river mountains 112
Wrangellia terrane 90, 100, 106, 136
Wyoming blocks 119

Yakutat terrane 74, 186
Yavapai 117
Yellowstone National Park 18
Yilgarn craton 5
Yukon river 181
Yukon-Tanana terrane 89